ON
THE PRACTICE
OF
SAFETY

FRED A. MANUELE, CSP, PE

 VAN NOSTRAND REINHOLD
_____ New York

Copyright © 1993 by Van Nostrand Reinhold
Library of Congress Catalog Card Number 92-32859
ISBN 0-442-01401-5

I(T)P   Van Nostrand Reinhold is an International Thomson Publishing company.
        ITP logo is a trademark under license.

Printed in the United States of America

Van Nostrand Reinhold   International Thomson Publishing GmbH
115 Fifth Avenue   Königswinterer Str. 418
New York, NY 10003   53227 Bonn
                                               Germany

International Thomson Publishing   _|P   International Thomson Publishing Asia
Berkshire House,168-173   221 Henderson Bldg. #05-10
High Holborn, London WC1V 7AA   Singapore 0315
England

Thomas Nelson Australia   International Thomson Publishing Japan
102 Dodds Street   Kyowa Building, 3F
South Melbourne 3205   2-2-1 Hirakawacho
Victoria, Australia   Chiyoda-Ku, Tokyo 102
                                               Japan

Nelson Canada
1120 Birchmount Road
Scarborough, Ontario
M1K 5G4, Canada

16 15 14 13 12 11 10 9 8 7 6 5 4 3 2

Library of Congress Cataloging-in-Publication Data
Manuele, Fred A.
        On the practice of safety / Fred A. Manuele.
            p.   cm.
        Includes index.
        ISBN 0-442-01401-5
        1. Industrial safety. I. Title.
    T55.M353   1993
    363.11—dc20   92-32859
                                         CIP

TO
IRENE

Contents

Preface

Throughout a rewarding career as a safety professional, I have observed with pride the achievements of the many who have contributed to the recognition of the practice of safety as a profession. We've come a long way. It is my hope that through this book I can assist in furthering that progress.

Ours is a profession in transition, with many additional opportunities available for involvement, accomplishment, and recognition. In light of the transitions in progress, the essays that comprise this book propose an extension of knowledge and participation by those engaged in the practice of safety, through which they can enhance their careers, become more effective, and be perceived as greater contributors toward achieving organizational goals.

This is also a book about fundamental principles and practices. It is addressed to students, educators, and practicing safety professionals. I hope that readers will welcome a review of our fundamentals, intended to improve the theoretical and practical base for our practice. I have attempted to argue systematically and consistently for hazards management ideas of substance in relation to our history, opportunities, and needs.

I expect that these essays will be slightly controversial since some long-accepted concepts are put to question. Some practices that have become fashionable are not sound. I understand that what I propose will be perceived by some to be heresy. Some readers will presume that I have put their finest ideals under scrutiny and proposed significantly different concepts.

In an atmosphere of quiet deliberation, my best hope is that what I have written will be perceived as thought-provoking and will serve to further the professional practice of safey.

Acknowledgments

To properly recognize all who have contributed to this book, I would have to cover a period of sixty-eight years and the list of people deserving appreciation would be endless. Particularly, though, to all the people who gave of their time in the past year and critiqued individual essays, I express my sincere thanks and gratitude.

Introduction

Since the essays in this book cover a broad range of subjects pertaining to the practice of safety, I thought it would be helpful to readers—students, educators, or safety professionals—if I provided guidance on their contents through a brief synopsis of each of them.

1. SAFETY PROFESSIONALS: THINGS ARE HAPPENING

Having perceived that important changes were taking place in the practice of safety, I also asked fifteen safety professionals for their views on the transitions that could have an impact on what we do, looking ahead five to ten years. This essay is a summary of that exercise, from which two conclusions can be drawn. Safety professionals can take the initiative, anticipate the changes coming in the practice of safety, become more effective as technical and managerial participants in support of attaining organizational goals, and be perceived as valued members of management staffs. Or they can ignore what is happening, and be demoted to technician status.

2. INCIDENT INVESTIGATION: A STUDY OF QUALITY AND A BASIS FOR IMPROVEMENT

Hazard analysis is the most important safety process. If that process fails, all other processes are likely to be ineffective. Incident investigation serves as one basis for hazard analysis. A collection was made from eleven companies of 537 actual incident investigation reports for a study of their quality. This treatise gives the findings of the study, provides a self-evaluation outline through which the quality of incident investigation can be assessed, and comments on how incident investigation can be improved.

3. ERGONOMICS: ITS IMPACT ON PROFESSIONAL SAFETY PRACTICE

It's estimated that 50 percent of work injuries and 60 percent of total workers compensation costs are ergonomics related. Skill requirements for safety professionals are in transition as ergonomics emerges to become a more significant element in safety and health program management. This essay looks to the future and explores the long-term impact of ergonomics on the content of safety practice. It will be significant.

4. SAFETY IN THE DESIGN PROCESS

It is my opinion that the greatest strides forward as respects safety, health, and the environment will be made in the design processes. This is where the first safety decisions are made. In a few organizations, safety professionals are so involved. Discussions include why safety professionals should become involved in design processes, how to get there, and what skills are needed.

5. SYSTEM SAFETY AND THE GENERALIST IN SAFETY PRACTICE

This essay follows in purpose the preceding treatise. As opportunities arise for generalist safety professionals to participate in design processes, an awareness will develop of the need for system safety skills. Most system safety literature is written in governmental jargon. It is the intent of this essay to establish why it is important for generalist safety professionals to acquire knowledge of system safety principles, to outline the system safety idea in nongovernmental terms, and to review briefly system safety methodologies.

6. ON THE PRACTICE OF SAFETY
AND TOTAL QUALITY MANAGEMENT

An examination of selected literature indicates that the management methods to achieve successful quality assurance are identical with those for successful safety management. Reviews are given of the criteria for the Malcolm Baldrige National Quality Award, principles espoused by W. Edwards Deming in Out of the Crisis, and the quality management program proposed by Philip B. Crosby in Quality is Free. Comments are also made about the quality management programs of two Baldrige Award winners. Safety professionals are encouraged to take the initiative and become members of quality management teams.

7. ON SAFETY, HEALTH, AND ENVIRONMENTAL AUDITS

I now believe that most safety, health, and environmental audits, intended to measure the quality of hazards management, are deficient for certain purposes. A self-review outline is presented that addresses the purposes of audits and speaks to what I believe to be the shortcomings in typical audit practices.

8. DEFINING THE PRACTICE OF SAFETY

We who call ourselves safety professionals will never be accepted as a profession until we agree on, make known, and meet the requirements of a definition of our practice. This essay identifies the societal need fulfilled by safety professionals, establishes why safety professionals exist, and defines the practice of safety.

9. ACADEMIC AND SKILL REQUIREMENTS
FOR THE PRACTICE OF SAFETY

Reviews are given of the academic knowledge and skill that would prepare one to enter the practice of safety, and of the knowledge and skill requirements that describe the applied practice of safety, using the term broadly.

10. ON BECOMING A PROFESSION

Safety practitioners will attain professional recognition when safety practice meets the regimens of a profession. Recognizing the great accomplishments thus far toward achieving professional recognition, a review of our present status, needs, and actions that we should consider are encompassed in a discussion outline.

11. ON CAUSATION MODELS FOR HAZARDS RELATED INCIDENTS (HAZRINS)

If several safety professionals investigate a given hazards-related incident, they should identify the same causal factors, with minimum variation. That is unlikely if the thought processes used have greatly different foundations. At least twenty causation models have been published. Since many of them conflict, all cannot be valid. A review of some of them is followed by a discussion of principles that should be contained in a causation model.

12. COMMENTS ON HAZARDS

Hazards are the justification for the existence of all safety professionals—whatever titles they use. Hazards are defined as the potential for harm or damage to people, property, or the environment. Hazards include the characteristics of things and the actions and inactions of people. Looking to the future, safety professionals will need a better understanding of the nature of hazards, especially for their participation in design processes. That understanding should include knowledge of Haddon's energy release concepts and the principles on which the MORT Safety Assurance System is established.

13. COMMENTS ON RISK

All risks with which safety professionals deal derive from hazards; there are no exceptions. Risk is a much-used term, with many meanings. In the context of the work of safety professionals, risk requires a measure of the probability and of the severity of adverse results. Thus two distinct aspects of risk must be considered by safety professionals in fulfilling their responsibilities: avoiding, eliminating, or reducing the probability of a hazard being realized; and minimizing the severity of adverse results if the hazard is realized.

14. ON HAZARD ANALYSIS AND RISK ASSESSMENT

Safety professionals cannot properly give advice on multiple hazards unless they are analyzed and categorized as to their potential and the possible severity of consequences, should their potentials be realized. This essay works through the process of hazard analysis and subsequent risk assessment. It explores hazard analysis and risk assessment methods, discusses the jeopardy in using some published techniques, and concludes with a practical methodology outline.

15. SUCCESSFUL SAFETY MANAGEMENT: A REFLECTION OF AN ORGANIZATION'S CULTURE

An organization's culture determines the probability of success of its hazards management endeavors. What the board of directors or senior management decides is acceptable for the prevention and control of hazards is a reflection of its culture. That theme is established. Also, interviews were conducted in organizations where the culture requires highly successful safety performance. Results of those interviews are set forth, along with comments on how safety management programs are made to work successfully where the culture so requires.

16. ANTICIPATING OSHA'S STANDARD FOR SAFETY AND HEALTH PROGRAM MANAGEMENT

It is my view that in the next few years OSHA will issue a Standard for Safety and Health Program Management. Safety and health professionals would do well to appreciate the possible outcome of current legislative action and what OSHA has issued so far—the requirements to participate in OSHA's Voluntary Protection Programs, Safety and Health Program Management Guidelines, and a Standard for Process Safety Management of Highly Hazardous Chemicals. From those issuances, a model has been extracted that probably will be close to what OSHA eventually proposes as a Standard.

17. ON MANAGEMENT FADS

For as long I can remember, there has been a continuing emergence of new management fads. There is a never-ending search for a magic pill to solve management problems. Fads don't last very long. But you can be sure that as one disappears, another will replace it, and safety professionals will be caught up in the cycle. Taking the long view, safety professionals should develop a constancy of purpose (a Deming phrase), no matter what fad is in vogue. A review is given in this essay of important works by Abraham H. Maslow and Douglas McGregor, which are the bases of many management innovations. Also, comments are made on some of the management fads of the last forty years.

ON
THE PRACTICE
OF
SAFETY

Chapter 1
Safety Professionals: Things Are Happening

Safety professionals can take the initiative, anticipate the changes coming in the practice of safety, become more effective as technical and managerial participants in support of attaining organizational goals, and be perceived as valued members of management staffs. Or safety professionals can ignore what is happening and be demoted to the status of a technician. These two statements are an appropriate summary for an inquiry concerning the transitions that could occur in the practice of safety, looking ahead five to ten years.

Safety professionals do not have a history of anticipating societal needs and of taking the leadership in fulfilling those needs. Typically, we are reactors.

What's evolving expands the practice of safety considerably. If safety professionals don't get into what's happening, others will be giving the advice that needs to be given, and management will perceive the roles those advice-givers play as being important. If safety professionals do only that which remains, they will be candidates for technician status.

It has been said that the half-life of a new engineer is only seven years. If that also applies to safety professionals, obviously a continuing education program is necessary to maintain a proper level of professional practice.

It is my view that OSHA's Standard for Process Safety Management of Highly Hazardous Chemicals (1) will have a long-term impact, far beyond the chemical industry. It is a safety management Standard. Safety professionals, whatever their endeavors, should be aware of its content and its possible impact on the practice of safety.

Now, a prediction. Within the next few years, OSHA will issue a Standard for Safety and Health Program Management. A framework in considerable detail for that Standard already exists. It is contained in the Safety and Health Program Management Guidelines issued by OSHA in 1989 (2), in the Star Program that is a part of OSHA's Voluntary Protection Program (3), and in the recently promulgated Standard for "Chemicals." Each of these issuances requires superior and demonstrated management commitment, involvement, and systems of accountability.

Promulgation by OSHA of a Standard for Safety and Health Program Management will, of itself, have significant impact. It will have a bearing on what may be expected of safety professionals in the future by senior managements concerning the quality of the technical and managerial content of their practices.

One highly significant aspect of the management systems required by OSHA in the three publications cited involves hazard analysis, both of existing facilities and of "planned and new facilities, processes, materials, and equipment." Having hazard analysis skills will become a necessity for safety professionals. For new facilities and for alterations, the time and place to anticipate, identify, eliminate, or control hazards is in the design phase. That presents great opportunity for upstream involvement by safety professionals. If they don't do it, someone else will.

Safety professionals becoming involved in facilities or equipment design from concept stage to conclusion fits in well with an evolving practice called "concurrent engineering"(4). This means simultaneously, not in sequence, examining the concerns of design, manufacturing, production, quality control, safety, purchasing, finance, and marketing—in the first stages of product and production process planning. Its theme is: do it right the first time.

A few safety professionals are thus involved, with great professional satisfaction. If, where concurrent engineering becomes the practice, personnel other than safety professionals are perceived as the source of advice on safety, particularly on safety in production processes, the status of the safety professional is diminished.

Also, note the references in the preceding to "process" or "processes." Identifying with the processes for getting things done will become more significant for safety professionals. Lester C. Thurow states in *From Head to Head: Coming Economic Battles Among Japan, Europe, and America:*

> In the twenty-first century, sustainable competitive advantage will come not from new-product technologies but from new-process technologies—those that enable industries to produce goods and services faster, cheaper, and better.(5)"

Safety professionals should pay attention to Thurow's prediction. This same emphasis on the "process" also appears in much of the literature on quality assurance.

According to W. Edwards Deming, a recognized leader in quality assurance:

> Quality comes not from inspection but from improvement of the process. The old way: inspect bad quality out. The new way: build good quality in. . . . Quality . . . must be built in at the design stage . . . and teamwork is essential in the process. . . . Redesign . . . and then again bring the process under control.(6)

Rummler and Brache, in *Improving Performance: How to Manage the White Space on the Organization Chart,* give this definition of process management: "a tool for improving organization performance through the identification and removal of cross-functional barriers to effectiveness." (8) Safety professionals have a role in "improving organization performance."

A review of several publications on quality assurance leads to an interesting conclusion. For paragraph after paragraph, the word *quality* can be replaced with *safety* and the premise remains sound. Methodologies to improve product quality are identical in most respects with those to achieve improvement in safety performance. Those "processes" that are a matter of concern to management personnel bent on improving quality are the same processes out of which injuries and illnesses occur. Is the work of the safety professional, as suggested by Dr. Thomas A. Selders, but another means of achieving quality performance?

Whether the subject is product design or production methods design or quality assurance, there is an opportunity here for the safety professional who wants to be perceived as assisting management to attain its goals. Get on the design review team and get on the quality assurance team!

Could there be a more noticeable overlapping of purposes with other organizational entities than is presented by the field of ergonomics?

Ergonomics is the art and science of designing the workplace and work methods to fit the worker. Ergonomics applies to safety, productivity, quality, and morale.

And, estimates frequently appear indicating that about 50 percent of work injuries and illnesses are ergonomics-related. Reducing the frequency of such injuries and illnesses requires modifications in workplace design and work methods design. Getting that done as a member of a team of which production and quality management personnel are also members just has to be more effective.

Surely, doing it right the first time, in the design stage, is preferable. But more frequent opportunities will arise for involvement with those who are responsible for the quality assurance analyses of the work processes in place. Resulting recommendations in work methods improvements should achieve lower injury and illness incident rates, as well as quality and productivity gains.

If safety professionals are not perceived as one of the sources of advice on ergonomics, recognizing that about 50 percent of work injuries and illnesses are ergonomics-related, the status of the safety professional is diminished.

To prepare for the completion of hazards analyses, participation in design concept discussions and design reviews, serving as a part of process review teams, and participating in quality improvement endeavors, some courses of study are recommended.

System safety concepts will apply throughout, and knowledge of them will be a requirement for professional safety performance. I believe that generalists in safety practice must acquire system safety skills.

For what seems to be evolving in the practice of safety, a refresher course in statistics would be beneficial. Several safety professionals have written about the applicability of quality assurance statistical methods to hazards management. To quote Deming, "Statistical thinking is critical to improvement of a system."

And how can a safety professional involved in occupational safety and health not be cognizant of ergonomics principles?

Another evolution is in progress concerning management in general, the substance of which I do not yet fully grasp. Certain terms are now common in the management literature: participative management, collaborative management, teamwork, commitment-based management, employee empowerment.

Fortunately or unfortunately, my management experiences have taken me through a multitude of management fads. They don't last very long; it seems that every few years, there is a new one. Whatever happened to such as management by committee, transactional analysis, intrapreneuring, and zero defects? How ironic that Deming says that zero defects, as a goal, makes no sense. Is it true that safety professionals are zero-defects driven?

Nevertheless, these new management approaches are intended to improve productivity and let workers make more decisions. If employees are truly sought for their input, their contributions to the practice of safety can be notable.

Having concluded that important things were happening, I asked fifteen safety professionals what they perceived to be on the horizon that would effect the practice of safety in the next five to ten years. Though their views vary considerably, each safety professional with whom I spoke predicted that substantive changes in the practice of safety would occur.

What follows is a fairly verbatim report of their responses, without editorial comments.

- Safety functions will be aligned with the engineering discipline, forcing nontechnical safety personnel into ancillary roles, as technicians. More companies will establish a key management position encompassing safety, health, and the environment, with one objective strategy and one source of advice and solution. It makes no sense, from an economic viewpoint and from a viewpoint of accomplishment, to look at only one aspect of a situation when there is an overlapping of safety, health, and environmental risks.

- A merging will occur of professional safety and industrial hygiene into a single practice, commencing at an academic level. Considering knowledge and scope requirements, a Master's degree may be required. Certification will become more important. Safety professionals may be behind the public demands.

- Safety professionals will have to sharpen both their technical and managerial skills. They should work to be a part of the management team, assisting in attaining management goals. Management, and society, will be expecting more of safety professionals, who can expect that opportunities to be a part of senior management staffs will be more frequent.

- Being a member of a Total Quality Management team allows the safety professional to help eliminate production problems and reduce incident frequency at the same time, and look very good in the eyes of management. "Fishbone" charts work equally well for both quality assurance and for hazards management. A course on Total Quality Management has demonstrated that the practice of safety is a natural fit with Total Quality Management.

- Regulatory burdens will increase, but they should not dominate the overall program. Demands of regulations can drive what a safety professional does. If all that is done is to comply with regulations, the organization doesn't have much of a safety and health program. Keeping the total content of a good safety and health program management in balance with complying with regulations will become a greater challenge. Being viewed solely as "the compliance person" can restrict promotional possibilities.

- Safety professionals should emerge as initiators, with solutions. They should be perceived as recognizing the needs and processes of a business and being a participant in improving the processes. Traditional approaches to management will give way to a more participative management. It's time now to train middle managers for the responsibilities they will have in five or ten years. A more educated population will ask tougher questions and have greater expectations.

- Responsibilities of more safety professionals will be extended, at least to include safety, health, and the environment. And there will be a greater attention to the design of work arising from such factors as stress illnesses and the Americans with Disabilities Act (8). Safety professionals will paint themselves into a smaller and smaller corner if they don't take the initiative as new needs arise.

- There will be a continued emphasis on human factors engineering, and on the management of process hazards. Safety professionals can provide a valuable link between engineering and operations. They must be more active in understanding management styles, in getting on the team, and must be perceived as problem solvers in relation to goals.

- Designed-in safety will become more prevalent, with safety professionals being a part of the design team. There will be more women in the ranks of safety professionals. They will be more sympathetic to the needs of minorities and women in the design of the workplace and in work methods design. As an example, ergonomics practice is not yet speaking to the abilities and limitations of minorities and women. Process review teams, consisting of design, quality control, manufacturing, human relations, and safety personnel, will be the way of the future.

- In the European Economic Community, the impact of safety regulation is steam-rolling. In the U.S., a major change will occur in corporate philosophy on safety, albeit slowly. Culture changes, with direction and accountability requirements coming from chief executive officers, will be more frequent. Safety professionals will have to become more professional, in both their technical and managerial capabilities.

- Because of costs, workers compensation will receive more executive attention, resulting in drives for improved injury and illness prevention, and for combating bogus claims. There will be a recognition that what has evolved for hazard communication is a sham and a means will be found to actually communicate chemical hazards information. Competitiveness will be a driving factor, giving more emphasis to ergonomics for both productivity and safety. Lessons learned will make teams created as a part of employee empowerment programs more effective.

- More executives will become aware that their safety needs cannot be met with the relatively few safety professionals they now employ. Some of this awareness comes from regulatory pressures. From the top to the bottom of organizations, there will be a greater demand for higher levels of safety performance. It has been proven that attaining both improved quality and safety can be done as a concerted effort, and that idea will spread.

- Safety engineering concepts will prevail. At the same time, a higher level of management skills will be needed. A greater push can be expected for design-in safety for occupational safety and health, for product liability prevention, and for environmental controls. Safety Sciences concepts now taught in Germany, which are principally engineering and science oriented, will have an influence in other countries. Safety professionals should be communicating, now, with management on what ought to be done to improve the quality of hazards management, the net effect being improved results for the organization. Those communications should be positive and show sufficient strength, without being offensive. Safety professionals will get there only by proving their value.

- There are two driving forces that will push safety professionals into more involvement with design and engineering. One is OSHA's long-stated position that engineering is the preferred solution for hazard prevention and control. And the other is ergonomics, which is workplace-design-oriented. Reporting relationships will change, the result being a combination of safety, health, and environmental affairs into one function under the direction of a senior management person.

- There will be a greater recognition of the importance of certification. Becoming a Certified Safety Professional will be a requirement for higher-level positions. Safety professionals will have to expand their horizons and become prepared for additional opportunities. When they are prepared, they must let management know that they can fulfill other roles.

Surely, those with whom discussions took place are predicting a changing world for the practice of safety. Their comments represent a great spread of views, with which one could agree or disagree. Some themes are repeated several times. Nevertheless, it can be reasonably assumed that the content of the practice of safety will be somewhat different in five to ten years than it is now.

These are but two of the positions safety professionals can take in light of the changes that could occur. They can stick their heads in the sand, ignore what is happening, bask in the status quo, and be demoted to the status of a technician. Or they can take the initiative, anticipate the changes coming in the practice of safety, become more effective as technical and managerial participants in support of attaining organizational goals, and be perceived as valued members of management staffs.

REFERENCES

1. *Process Safety Management of Highly Hazardous Chemicals.* OSHA Standard, 1910.119, February 1992.

2. *Safety and Health Program Management Guidelines.* OSHA. January 1989.

3. *Star Program.* OSHA Voluntary Protection Programs, July 1988.

4. John R. Hartly, *Concurrent Engineering.* Cambridge, Mass: Productivity, 1991.

5. Lester C. Thurow, *From Head to Head: Coming Economic Battles Among Japan, Europe, and America.* William Morrow and Company, 1992.

6. Mary Walton, *The Deming Management Method.* New York: Putnam Publishing Company, 1986.

7. Geary A. Rummler, and Alan P. Brache. *Improving Performance: How to Manage the White Space on the Organization Chart.* San Francisco: Jossey-Bass Publishers, 1991.

8. The Americans with Disabilities Act, July 1990.

Chapter 2

Incident Investigation: A Study of Quality and a Basis for Improvement

To study the quality of their contents, a collection was made of accident, injury, illness, mishap, and incident investigation reports completed by supervisors or investigation teams.

Why study the quality of incident investigation? Because, as stated in *MORT Safety Assurance Systems,* "Hazard analysis is the most important safety process in that, if that fails, all other processes are likely to be ineffective." (1) I emphatically agree with that premise.

Safety professionals exist for these purposes only:

- to anticipate, identify, and evaluate hazards, and

- to give advice on the avoidance, elimination, or control of hazards, to attain a state for which the risks are judged to be acceptable.

Application of that advice should result in the prevention or mitigation of harm to people, property, and the environment.

There are two major focuses in the work of safety professionals. One involves the anticipation, identification, and evaluation of hazards and the giving of advice concerning them *before* their potentials are realized as incidents. The second requires the identification and evaluation of the hazards that are the causal factors for incidents *after* they occur.

Thus, competent investigation of hazards-related incidents is vital to the professional practice of safety and to the success of a safety and health management program. Safety professionals who provided the reports collected for this study agreed with that premise.

To analyze the hazards that are the causal factors for the incidents that do occur, one of the methods used by safety professionals is to have investigation reports completed. That would typically be the responsibility of supervisors. In some organizations, if the results of incidents are serious, or if it is judged that results of an incident could have been serious under other circumstances, investigation teams would make a study in greater depth than would be expected of a supervisor.

Hazards identified in incident investigation reports should pertain to the actual causal factors if corrective action is to be properly directed. That requires a superior quality of incident investigation. One of the safety professionals who contributed to this project offered this view: "If you don't have confidence in the ability of supervisors to complete an incident investigation, you can't have much of a safety program."

I would like to emphasize that the study I made does not meet the modeling and methods requirements of scientific inquiry. From thirty-seven locations of eleven organizations, five hundred thirty-seven investigation reports, using fifteen different forms, were collected and reviewed. Quality of reports, generally, was assessed concerning:

- general information
- incident descriptions
- causal factors determinations
- corrective actions proposed

General observations deriving from the study follow. They are just that—observations—resulting from a subjective review of a limited number of reports—the 537 reports received. I do suggest that these observations deserve a broader consideration through a more scientific study since the subject—incident investigation—is such an important element in effective safety and health management programs.

These are the highlights of the study.

1. Using a scale of 1 to 10, with 10 being the best score for quality, individual entity scores ranged from a high of 8 to a low of 2. For the entire study, 5.5 was the weighted overall score.

2. Investigation of incidents is well done in those entities where the culture includes a management accountability for hazards-related results.

3. Variations in structure and content requirements in incident investigation forms were extensive.

4. Good incident investigation can not be achieved without training, and repeated training.

5. Where supervisors had participated in job hazard analyses, they seemed to have a better understanding of causal factors and did a more thorough job of incident investigation.

6. General information entries (name, social security number, occupation, et cetera), and incident descriptions were usually the most complete parts of reports.

7. Incident report forms are predominantly "Heinrichian": they emphasize "man failure"—a term significant in Heinrich's premises.

8. Although not necessarily intended, some report formats lead to identification of a single causal factor, emphasizing unsafe acts, rather than stressing the concept of multiple causation. Sometimes—only sometimes—a contributory causal factor is also identified.

9. It is probable that supervisors, who make only 1 or 2 or 3 incident investigations a year, do not fully understand the descriptions of the various causal categories given in investigation forms.

10. Form content requirements seldom lead to examinations of the design of the workplace or the design of work methods.

11. For approximately 38 percent of the reports reviewed, entries suggested that further inquiry should have been made as respects design of the workplace or design of work methods.

12. In the one instance reviewed in detail, cause codes entered for subsequent computer analysis did not match the contents of the investigation reports.

13. Injury type and incident type codes were close to reality.

14. For many reports, plausible causal factors could not be identified.

More detailed comments will now be made on the methodology and findings of the study.

Using subjective judgments, scorings were first given where possible for general information, incident descriptions, causal factors determinations, and corrective actions proposed. A score was given to the entire incident report. Subsequently, a composite score was then given to the entity's system as a whole. A quality score of 10 was best, on a 1 to 10 scale. All scores were rounded to whole numbers. Composite scores ranged from a high of 8 to a low of 2, with an overall weighted score of 5.5 for the entire study.

Some organizations are doing a very good job of incident investigation. Reports were commendable, proving that effective incident investigation can be achieved. In other companies, what is accepted indicates inadequate knowledge by those completing investigations and no management accountability for incident investigation. Completion of reports in these instances is obviously a perfunctory exercise, with little value.

COMPOSITE SCORES

	10	9	8	7	6	5	4	3	2	1
Individual			X	X	X	X		X	X	
Entities			X	X		X		X	X	
Overall Weighted Score				5.5						

Structure and content variations in incident investigation forms were extensive. It is not suggested that precise and equivalent evaluations could be made in reviews of reports. As an example, one of the simplest forms consisted in its entirety of these questions: What happened? Why did it happen? What should be done? What have you done so far? How will this improve operations?

Discussions of achievements with safety professionals whose organizations had top scores did not produce any surprises. Incident investigation for hazard identification and analysis gets done best where the organization's culture includes an accountability for superior performance in avoiding harm or damage to people, property, and the environment.

An aggregate follows of the comments made by safety professionals in those entities with the best incident investigation systems.

• A focus on safety in those entities comes from the top, as a reflection of the organization's culture.

• If incident investigation is important to the "boss," if it is a part of the accountability system, it will be well done. Where that is the case, requests for help will come from all levels of management.

• For everyone from the senior vice president down to the floor supervisor, the annual performance review includes safety performance.

• Safety performance is an element scored in the bonus program.

• There is an annual competition on safety performance. Statistics produced monthly are treated seriously.

• Only one award in the organization has the president's name on it—the safety achievement award. Presentation of the award is made by the president at a significant social event.

• For every recordable incident, the location manager is required to submit a report to the group vice president at headquarters.

• At all manufacturing and warehousing locations, there is a job hazard analysis for every job.

• Supervisory personnel can not do a good job of investigation if they have not had the necessary training. Thus, training programs are conducted on incident investigation, and repeated.

Reports have a variety of titles, although some were identical. They come under these names, which in themselves may have significance:

- Investigation and Findings of Injury/Illness
- Supervisors Report of Accident Investigation
- Mishap Report
- Accident Investigation Report
- Injury/Illness Report
- Incident Investigation Report: Personal Investigation
- Supervisor's Investigation Report
- Occupational Injuries or Illnesses Report

General information entries were usually well completed. A summary of entries required by all forms include:

Name	Social security number
Sex	Clock number
Shift/time	Marital status
Date of birth	Address
Department/division/sector	Occupation/job title
Seniority date/company service	Number of years/months on job
Date of accident/illness	To whom was accident reported
Where sent/hospital/home/other	Describe medical treatment
Did employee die	Basic cause code
Contributory cause code	Accident code
Injury/illness code	Place of accident
Was place of accident on employer's premises	

Incident descriptions were often the most complete parts of reports. In many instances, incident descriptions were the only complete parts of reports. It is easier, obviously, to describe what happened than it is to determine causal factors and what corrective actions should be taken.

Causal factor determinations varied greatly as to quality. Although three of the fifteen forms received make no reference to unsafe acts or unsafe conditions, causal identification requirements of incident investigation reports are still predominantly "Heinrichian."

After a review of his "theorems," developed in the 1920s and illustrated by the "domino sequence," H. W. Heinrich, in the third edition of *Industrial Accident Prevention,* wrote:

> From this sequence of steps in the occurrence of accidental injury it is apparent that man failure is the heart of the problem. Equally apparent is the conclusion that methods of control must be directed toward man failure.(2)

Of the fifteen forms received, ten just about—almost—direct the person who completes the form to first identify the unsafe act of the employee. That requires seeking evidence of the "man failure" to which Heinrich referred. It is a prominent practice, whether intended or not, to put the principal responsibility (not blame) for the incident on something the employee did or did not do. Yes, some of the forms also ask that unsafe conditions be recorded. Sometimes they contain references to design and systems shortcomings.

In those organizations where incident investigation is done well, supervisors who obviously had training on incident causation and were familiar with job hazard analysis procedures would frequently go beyond the form requirements. It was not unusual for those supervisors to pose questions about and seek help on the design of the workplace and on the design of work methods. Also, it regularly occurred in those organizations that supervisors would record on incident investigation reports something like "this job needs a new job hazard analysis."

Form structure and content seldom lead to examinations of workplace design or work methods design.

For approximately 38 percent of the reports reviewed, entries suggested that further inquiry should have been made concerning design of the workplace or design of work methods. In far too many cases where recordings indicated that a review of workplace design or work methods design might be beneficial, unsafe acts of employees were selected as the primary incident causes. And the corrective actions proposed were to obtain behavior modification.

As a generality, few supervisors are yet qualified to do a good job of ergonomics problem identification. Although there has been much emphasis in recent years on ergonomics as a significant aspect within safety and health program management, that emphasis has not influenced the content of the incident investigation reports I reviewed.

Terms such as basic cause, contributory cause, immediate cause, and management cause appear on the forms. And it is assumed that they are understood by the safety professionals who put them there. But definitions of those terms may not convey to supervisors what is intended. Entries on investigation forms raise questions concerning how well supervisors, who may complete 1 or 2 or 3 incident investigation reports a year, understand what the terms mean. Definitions given on the forms were understandably brief and were sometimes confusing. I was not always certain of their meanings.

In five of the fifteen forms received, causal factor, incident, and injury type codings are required. This is a subject for which I had been an aggressive promoter—the use of causal factors codes for later computer entry and summary. Now, I would pose a question about the accuracy of causal data that can be derived from investigation reports completed by supervisors.

For a group of 121 reports, these basic cause codes were entered in the boxes for such codes, all of them employee action related.

28%	Failure to follow established procedure
27%	Haste, inattention, shortcut
_6%_	Improper use of equipment, tools, or materials
61%	

Basic cause codes selected did not match the written entries on the forms. Supervisors were doing a better job of identifying possible causal factors than the codings would indicate. Corrective actions taken were also in greater depth than would be required by the cause codes chosen. Computer runs would subsequently identify principle causal factors based on those cause codes. Using that data as factual would result in a misdirection of efforts.

Safety professionals who have systems requiring the entry of cause codes on investigation reports for which computer summaries are produced might want to conduct a similar exercise, to determine whether the computer output matches reality. I suggest a close look at what is being obtained, and an assessment of what can realistically be obtained in a given setting.

For many reports, plausible causal factors could not be identified. Significant shortcomings in incident investigation for causal factor determination were obvious. Yet, to repeat, incident investigation and subsequent hazards analyses are such vital aspects of safety and health program management.

Of the fifteen variations of incident investigation reports received, six promote an overly simplistic and inappropriate approach to causal factor determination. They reflect an instruction, although somewhat ancient, given in a publication of the American National Standards Institute:

> It is recognized that the occurrence of an injury frequently is the culmination of a sequence of related events, and that a variety of conditions or circumstances may contribute to the occurrence of a single accident. A record of all these items unquestionably would be useful to the accident preventionist.

> Any attempt to include all subsidiary or related facts about each accident in the statistical record, however, would complicate the procedure to the point of impracticality. The procedure, therefore, provides for recording only one pertinent fact about each accident in each of the specific categories or classifications. (3)

Incident causation can be complex. Yet there seems to be a desire to retain an age-old theme of simplicity. If people who make incident investigations are directed to select "one pertinent fact . . . in each of the specific categories"—"*the* unsafe act" or "the unsafe condition"—more than likely that is what they will do, thus diminishing the value of the investigation. Where incident investigation is done best, multiple causal factors are sought and it is the exception when only a single causal factor is recorded.

In some entities, it is required that an incident investigation team be selected and gathered if the results of the incident were serious or if the results of the incident could have been serious under other circumstances. Those reports—twelve of them were received—were a pleasure to read. Every one reflected an understanding of multiple causation and pursued several routes in causal factors determination and in selecting corrective actions.

Injury type and incident type codings were at a much higher accuracy level than codes for causal factors. Incident type and injury type analyses would provide information from which to commence further inquiry, but that's all.

It is a too-common practice that personnel with "safety" in their titles place their signatures, indicating acceptance, on incident investigation reports that are far from adequate.

Although "employee empowerment" has been prominent in recently published management literature, there were no indications in this study that personnel other than supervisors are completing incident investigation reports.

Surprisingly, the terms "careless" or "carelessness" or "should have been more careful" appeared only seven times in 537 reports.

It is my belief that a score of 5.5 for incident investigation, the weighted overall score for this study, should be considered inadequate by safety professionals for this very important function.

How would safety professionals benefit from this study? How would safety professionals go about improving the quality of incident investigation?

First, I propose that they review the findings of this study in relation to their own incident investigation systems. Then I propose a self-audit of the quality of incident investigation in the operations for which they have responsibility. Determining how well incident investigation is performed requires, first, some introspection to establish the concepts of incident causation that the safety professional believes should apply.

Assume that a decision is made that Heinrich's principles of incident causation, although antiquated, are to be the bases for the design of incident investigation reports and for the measurement of the quality of incident investigations.

Or assume that Heinrich's principles are considered inadequate and that the causation model should include an understanding of Haddon's energy release theory (4), an awareness of the concepts on which the Management Oversight and Risk Tree (MORT) (1) is based, a comprehension of system safety concepts, and an appreciation of the implications of the design of the workplace and the design of work methods as causal factors.

Content of the self-audit would be markedly different, depending on the assumptions made. Nevertheless, an attempt will be made at producing a self-audit outline.

QUALITY OF INCIDENT INVESTIGATION SELF-AUDIT OUTLINE

1. First, a safety professional must determine what incident causation model is to apply. All that follows in this outline regarding causal factors will relate to that model.

2. Taking a sample of incident investigation reports completed by supervisors, rate the following for quality:

   a. General information data

   b. Incident descriptions

   c. Recording of injury and illness data

   d. Completion of code entry requirements

   e. Causal factors determinations

   f. Actions to be taken to prevent recurrence

   g. Completion of questions particular to the organization

   h. General overall quality of the report

3. Using the same sampling of reports, make these assessments:

   a. Are the terms on the forms understood?

   b. Are causal entries on forms too simplistic?

   c. Do causal entries fit with your concepts of incident causation?

   d. Does the quality of the reports show a need for additional training in incident investigation?

   e. Do the contents of forms indicate a need for additional job hazard analyses?

   f. Does the data gathered allow meaningful analyses?

   g. Is the needed managerial attention given to reports?

   h. Are safety personnel fulfilling their responsibilities?

4. Concerning an incident investigation procedure or guideline, and the incident investigation report:

   a. Do they clearly convey an understanding of the causation model that you want applied?

   b. Are the definitions clear? Is the language ambiguous?

   c. Do they properly balance consideration of engineering and design, the work environment, equipment, work methods, employee behavior aspects, and management systems?

   d. Do they clearly establish which incidents are to be investigated?
      - Personal injuries
      - Significant property damage
      - Incidents not resulting in severe injury or significant property damage but that could have in other circumstances

   e. Are responsibilities for investigation clearly defined?
      - Supervisors
      - Upper management
      - When incident teams are to be used
      - Safety personnel

   f. For report distribution, are the approval levels what they should be?

   g. For corrective action
      - Are responsibilities precisely established?
      - Is the follow-up procedure appropriate?

   h. Do the procedures or guidelines and the incident investigation form need revision and re-issuance?

5. Concerning training in incident investigation:

   a. Should the program be improved?

   b. Has it been given to all who need training?

   c. Are refresher courses given?

   d. Is an additional focus on training needed?

6. Is there a need to convince management that giving greater significance to quality incident investigation will assist in their achieving their goals?

7. Having made this self-audit, what actions are to be taken?

It would be folly to suggest that a single incident investigation system could be drafted that would universally apply. I do propose that in determining what incident investigation system is to be adopted, an assessment be made of that which is practicably attainable.

That requires making assumptions about the organization in which the safety professional resides. If there is little management support for incident investigation or if training can not be given to those who are to make investigations, it would be best to opt for simplicity.

Whatever the situation, safety professionals should be trying to improve the quality of this important aspect of safety and health program management. And those improvements should reflect what has been learned.

For instance, incident investigation systems should be directing inquiry very early on into what may have been "programmed" through the design of the workplace or design of work methods. They should be promoting inquiry that determines whether the design of the work methods are "error provocative." This has become particularly obvious with the emergence of the greater significance of ergonomics as a segment of safety practice. For all musculo-skeletal injuries, and for many others, the first question should be: Are there workplace design or work methods design implications?

To extend that idea, this example is quoted from the chapter titled "The Error-Provocative Situation: A Central Measurement Problem" in *Human Factors Engineering* by Alphonse Chapanis. It is contained in the book titled *Measurement of Safety Performance* by William E. Tarrants.

> [S]ix infants had died in the maternity ward . . . because they had been fed formulas prepared with salt instead of sugar. The

error was traced to a practical nurse who had inadvertently filled a sugar container with salt from one of two identical, shiny, 20-gallon containers standing side by side, under a low shelf in dim light, in the hospital's main kitchen. A small paper tag pasted to the lid of one container bore the word "Sugar" in plain hand-writing. The tag on the other lid was torn, but one could make out the letters "S..lt" on the fragments that remained. As one hospital board member put it, "Maybe that girl did mistake salt for sugar, but if so, we set her up for it just as surely as if we'd set a trap. (5)

In the study of the quality of incident investigation just completed, 38 percent of entries suggested that further inquiry should have been made into the design of the workplace or of work methods. In far too many cases, the incident occurred in an "error-provocative situation." Yet in those instances, only unsafe acts of employees would usually be recorded as the incident causes.

Dr. Chapanis offered four axioms that deserve thought both in determining what incident causation model is to apply and in drafting an incident investigation report form:

Axiom 1. Accidents are multiply determined. Any particular accident can be characterized by the combined existence of a number of events or coincidence of a number of events or circumstances.

Axiom 2. Given a population of human beings with known characteristics, it is possible to design tools, appliances, and equipment that best match their capacities, limitations, weaknesses.

Axiom 3. The improvements in system performance that can be realized from the redesign of equipment is usually greater than the gains that can be realized from the selection and training of personnel.

Axiom 4. For purposes of man-machine design there is no essential difference between an error and an accident. The important thing is that both an error and an accident identify a troublesome situation.

Very few safety texts treat incident investigation in depth. For reference purposes the following are recommended.

- *Investigating Accidents With STEP* (6), by Kingsley Hendrick and Ludwig Benner, Jr.

  According to the preface, "The heart of the book is its presentation of the Sequentially Timed Events Plotting (STEP)." Also, "Building on system safety technology and the safety assurance systems of the Management Oversight and Risk Tree (MORT), the accident investigation methodology presented relies on a new conceptual framework."

- *MORT Safety Assurance Systems* (1) by William G. Johnson.

  This serves well both for incident causation model building and for incident investigation. The "Accident Investigation" chapter states, "Accident investigation has always been a major element in safety. Pre-accident hazard analysis is preferable, of course."

  MORT as an accident investigation and analysis technique promotes a thorough inquiry into the multiplicity of causal factors, of an engineering and technology nature, and of a management systems and behavioral nature.

  It could be appropriately argued that MORT in its entirety is a bit much for the simple and ordinary and that a simplified version would be more suitable for use by other than safety professionals. Nevertheless, MORT offers a sound thought process upon which to build an incident investigation system.

- *Modern Accident Investigation and Analysis: An Executive Guide* (7) by Ted S. Ferry.

  This text helps in thinking about how incidents occur and how they should be investigated:

  > Accident prevention depends to a large degree on lessons learned from accident investigation.

  > We cannot argue with the thought that when an operator commits an unsafe act, leading to a mishap, there is an element of human or operator error. We are, however, decades past the place where we stopped there in our search for causes.

  > While the traditionalist will seek unsafe acts or unsafe conditions, the systems person will look at what went wrong with the system,

perceiving something wrong with the system operation or organization that allowed the mishap to take place.

Ferry wrote extensively about System Safety, Change Analysis, the MORT Process, Multilinear Events Sequencing, and the Technic of Operations Review (TOR) (8), for which D. A. Weaver was the author.

- *Accident Investigation . . . A New Approach* (9), National Safety Council.

This forty-four page publication is highly recommended as an excellent reference for those who would evaluate and improve their incident investigation systems. There are nine references in its bibliography, six of which are mentioned in this paper: ANSI, Benner, Ferry, Johnson, Weaver, and the National Safety Council. Its focus is on causal factors. No mention is made of unsafe acts or unsafe conditions.

Its centerpiece, "A Guide for Identifying Causal Factors and Corrective Actions," requires a systematic approach to identifying causal factors and selecting corrective actions. It promotes consideration of multiple causal factors, rather than stopping with the typical "unsafe act." It also provides an additional thought base when considering a training program.

There are four major sections in the guide: Equipment, Environment, People, and Management. For each category, there is a listing of questions under the caption "Causal Factors" and, opposite the questions, are "Possible Corrective Actions."

The following two abbreviated examples are from the guide.

Causal Factors	*Possible Corrective Actions*
Did the design of the equipment/tool(s) create operator stress or encourage operator error?	Review human factors engineering principles. Alter equipment/ tool(s) to make it more compatible with human capability and limitations. . . . Encourage employees to report potential hazardous conditions created by equipment design.

| Were any tasks in the job too difficult to perform (for example, excessive concentration or physical demands)? | Change job design and procedures. |

In a form appropriate to the organization in which a safety professional is domiciled, a system of this sort should result in focusing on actual causal factors and on relative corrective actions. For those who seldom make incident investigations, a modification of the "Guide for Identifying Causal Factors and Corrective Actions" would serve as a valuable memory jogger.

I believe that the state of the art in this aspect of the practice of safety—incident investigation—will be moved forward through broader application by safety professionals of:

- Haddon's energy release theory,
- the concepts embodied in the Management Oversight and Risk Tree (MORT),
- the principles of multiple causation and of sequentially timed events plotting,
- system safety concepts, and
- the premise that the first inquiries as to causal factors for an incident should concern design and engineering—determining whether the workplace or work practices prescribed were "error provocative."

To improve the quality of incident investigation, a safety professional would determine the causation model to be applied, conduct a self-audit of the quality of incident investigation, and identify and implement the system improvements necessary. I hope that this essay is of value in such an exercise.

REFERENCES

1. William G. Johnson. *MORT Safety Assurance Systems.* New York: Marcel Dekker, 1980.

2. H. W. Heinrich. *Industrial Accident Prevention,* 3rd ed. New York: McGraw-Hill, 1950.

3. *Method of Recording Basic Facts Relating to the Nature and Occurrence of Work Injuries.* American National Standards Institute, New York, publication Z16.2, item 2.5.

4. William Haddon, Jr. "On the Escape of Tigers." *Technology Review,* May 1970.

5. William E. Tarrants. ed. *The Measurement of Safety Performance.* New York: Garland Publishing, 1980.

6. Kingsley Hendrick and Ludwig Benner, Jr. *Investigating Accidents with STEP.* New York: Marcel Dekker, 1987.

7. Ted S. Ferry. *Modern Accident Investigation and Analysis: An Executive Guide.* New York: John Wiley & Sons, 1981.

8. D. A. Weaver. "Technic of Operations Review—TOR Analysis. *Journal of the American Society of Safety Engineers,* June 1973.

9. *Accident Investigation . . . A New Approach.* Itasca, Ill.: National Safety Council, 1983.

Chapter 3

Ergonomics: Its Impact on Professional Safety Practice

Ergonomics is defined as the art and science of designing the workplace and work methods to fit the worker. Its generic base is engineering.

In some places, that "art and science" has been called human factors engineering. Alphonse Chapanis, in his article "To Communicate the Human Factors Message, You Have to Know What the Message Is and How to Communicate It," wrote:

> Human factors engineering is the application of human factors information to the design of tools, machines, systems, tasks, jobs, and environments for safe, comfortable, and effective human use.(1)

Dr. Chapanis also expressed these views:

> I don't intend to enter into an extended discussion about the differences between human factors and ergonomics. Frankly, I don't think the differences, such as they are, are important. . . . So though I shall be using the words human factors and human factors engineering in this article, I mean them to apply equally to ergonomics and the practice of ergonomics.

Ergonomics and human factors engineering have become synonymous terms. In this essay, ergonomics also means human factors engineering.

What is the significance of ergonomics-related incidents within the universe of workplace injuries and illnesses?

At the Western Safety Conference held in Anaheim in 1991, Gerard F. Scannell, then Assistant Secretary of Labor for Occupational Safety and Health, said that it is estimated that 50 percent of OSHA reported cases are ergonomics-related. That approximation ties in closely with the results of a study made by a major insurance company. Working with a claims cost base of more than $1 billion, it was concluded that about 50 percent of workers compensation injuries and illnesses and 60 percent of total costs were ergonomics-related (2).

A study made by Towers, Perrin, Forster & Crosby, Inc. (3) showed that the direct workers compensation costs in 1990 for job-related injuries and illnesses would be more than $60 billion. Then, the direct workers compensation costs—not the total costs—of ergonomics-related injuries and illnesses in the United States in 1990 could have been $36 billion.

Discussions with several safety directors resulted in general agreement with these estimates of the significance of ergonomics within the universe of worker injuries and illnesses. Two cautions were expressed. Estimates are probably applicable if the statistical sample is large enough. And variations by industry could be significant.

A professor in an industrial engineering department who is responsible for courses on ergonomics was asked to comment on these numbers. He said they were close to his assumptions.

Unfortunately, ergonomics is narrowly perceived by some to include only cumulative trauma disorders. Ergonomics concepts should be applied to every aspect of work station and work methods design that could present excessive biomechanical stresses, either cumulative or instantaneous.

Back injuries provide an illustrative case in that respect. In determining causal factors for back injuries, whether cumulative or instantaneous, basic questions concerning the possible implications of work station and work methods design should be asked. Not doing so would result in opportunities being missed for substantive reduction in the number of back injuries.

Extend that thinking to include all incidents involving overexertion and the statistical base of injuries and illnesses that may be ergonomics-related is broadened, far beyond just cumulative trauma disorders.

In the 1991 edition of *Accident Facts* (4), published by the National Safety Council, overexertion is said to represent 31.3 percent of all disabling work injuries for 1988. A summary of estimates obtained from three large insurance companies indicates that about 25 percent of all reported workers compensation claims involve overexertion. Estimates of payments for those claims range from 33 percent to 40 percent of total claims costs.

Assume that, on a macro basis, estimates previously cited are close. On a best-information-available-basis, it will be accepted that 50 percent of total worker injuries and illnesses and 60 percent of total workers compensation costs are ergonomics related. What introspection do these estimates suggest?

General awareness of ergonomics as a significant factor in determining what risk reduction measures should be taken is a recent development. Historically, application of ergonomics principles were seldom a part of professional safety practice. From those observations, these questions follow:

- Why haven't our analyses of workplace incidents revealed the significance of workplace and work practice design as causal factors?

- Are our incident analyses so shallow that the remedial actions we derive from them are often wrong?

- Does the well-used phrase "garbage in, garbage out" apply to most incident causal analyses?

- Have the remedial actions we have proposed really addressed the basic causal factors?

- How much of the time spent by safety professionals on occupational safety and health should be devoted to ergonomics?

It is the job of safety professionals to give advice on the avoidance, elimination, or control of hazards to prevent or mitigate harm or damage to people, property, and the environment. In fulfilling that role, the professional practice of safety fundamentally requires that, first, an accurate job be done of anticipating, identifying, and evaluating hazards. Otherwise, the advice given to avoid, eliminate, or control hazards will not relate to actual causal factors.

Hazard analysis is the first step in safety and health program management. Do that poorly, and all that follows is ineffective.

Hazards are hazards. They are the justification for the existence of all safety and health program activities. And ergonomics hazards fall within the spectrum of hazards to be dealt with in achieving safety and health program effectiveness.

If it is a close estimate that 50 percent of workplace injuries and illnesses are ergonomics-related, the professional practice of safety requires extensive involvement in ergonomics by those who have responsibilities in occupational safety and health. If a safety practitioner is to give advice on preventing or mitigating harm to employees and that person is not deeply into ergonomics, this question is appropriate. What does that safety professional do, and how does that activity relate to real needs?

Recognition of ergonomics as a requirement for successful safety and health program management has already had a significant impact on the content of professional safety practice. Looking to the future, what other long-term developments can be expected?

- A greater awareness will evolve of the favorable results to be gained, both for risk reduction purposes and productivity, by anticipating and avoiding hazards in the design process.

- Anticipation of hazards in the design process will consider not only the physical aspects of the work station but will also emphasize the possible stresses of the work methods to be prescribed.

- More safety professionals will become prepared to give counsel in the design process. Having demonstrated success, they will be sought for that purpose.

- Additional methods analysis skills will be developed by safety professionals, resulting in more accurate determinations of incident causal factors.

- Incident investigation and analysis will commence with these questions: Are there implications of workplace design? Are there implications of work methods design? Is the design of the workplace or the work methods error provocative?

- Inspection systems will be reoriented to include a greater attention to possible incident causal factors deriving from workplace and work methods design.

- Safety audit systems will acquire a new, major element: safety in the design process.

- Training programs, safe procedure guides, agendas for safety committee meetings, health hazards controls, hazardous materials controls, and maintenance methods will be revised to include ergonomics considerations.

- An awareness will re-emerge that, in order of preference, the most effective hazards management technique is to design hazards out of the work station and out of work methods.

Modifications that have come about in the investigation and analysis of manual material handling incidents are a good example of the impact that ergonomics is having on the practice of safety. Reviews of analyses made in the past by some safety professionals resulted in the conclusion that, often, when "improper lifting" was given as the incident cause, the root cause was actually workplace design. "Improper lifting", so called, had been programmed into the work procedure.

Consider this example. Bags weighing 100 pounds are delivered to work stations on a pallet that is set on the floor. Workers slit open the bags and lift them to shoulder height to pour the contents into hoppers. Speed, stooping, and twisting are required. Back injuries are reported. Causes in investigation reports are always recorded as improper lifting. Corrective action is always reinstructing the worker in proper lifting techniques.

This is obviously a work station and work method design problem. No amount of training or reinstruction or behavior modification could be an adequate solution. Yet safety professionals accept that this scenario relates too often to the actuality of past investigation and analysis practices. For remedial action, another employee training program would be conducted on "how to lift safely." And back injuries would continue to occur.

I believe that the greatest strides forward as respects safety, health, and the environment will be made in the design processes. Several safety professionals who recently participated in design discussions—a new venture for them—have been euphoric when recounting their experiences. They expressed an added sense of accomplishment. That's understandable.

Without question, the content of professional safety practice is undergoing a necessary and vital transition. To prepare for the needs and opportunities presented by this evolution, what basic ergonomics-related knowledge and skill should be acquired by the safety professional?

- Working knowledge of engineering principles is necessary to communicate successfully with engineers who make work place and work practice design decisions.

- An understanding is requisite of the broad range of human physical and psychological variations that employees bring to the workplace, which is the subject of anthropometry.

- Knowledge of the application of mechanical principles for the analysis of forces on body parts will be necessary, and can be obtained through a study of biomechanics.

- Appreciation is needed of the shortcomings of designing a fixed and inflexible workplace to average characteristics, the all too typical design practice.

- Through the foregoing, knowledge and skill would be attained concerning what constitutes good work station and work methods design.

A great number of ergonomics courses are now available. Some are superficial. Others are in appropriate depth in relation to needs and opportunities. Considering the impact that ergonomics is having and will have on the content of professional safety practice, the courses of study taken should be substantive.

Acquiring an adequate library of ergonomics related texts has also become requisite. Following is a recommended listing chosen from many possibilities, moving from the basic and necessary to the more complex.

- *Work Practices Guide for Manual Lifting* (5), by the National Institute for Occupational Safety and Health (NIOSH).

  A safety professional interested in back injury prevention must be familiar with the contents of the NIOSH study on manual lifting.

- *Cumulative Trauma Disorders: A Manual for Musculoskeletal Diseases of the Upper Limbs* (6), edited by Vern Putz-Anderson.

  Putz-Anderson's book is an excellent primer and is a must as a reference. Contributors have been prominent in the development of ergonomics concepts. Its well-written approach to prevention is sound.

- *Ergonomics: A Practical Guide* (7), by the National Safety Council.

  The National Safety Council's ergonomics guide provides a basic outline that is helpful in getting started on an ergonomics program.

- *Ergonomic Design for People at Work,* volumes 1 (8) and 2 (9), by the Human Factors Section at Eastman Kodak.

  Both of these volumes are practical and easily followed.

- *Fitting the Task to the Man: An Ergonomic Approach* (10), by E. Grandjean.

  Considering the number of times Grandjean is quoted, this volume has obviously proven its value as a resource.

- *The Practice and Management of Industrial Ergonomics* (11), by David C. Alexander.

- *Fundamentals of Industrial Ergonomics* (12), by B. Mustafa Pulat.

  The Practice and Management of Industrial Ergonomics and *Fundamentals of Industrial Ergonomics* are used in college-level ergonomics courses in engineering curricula.

- *Human Factors Design Handbook* (13), by W. E. Woodson

  Woodson's book is what the title implies: a design handbook.

For all of these texts, a theme is apparent—design the job to fit the worker.

Safety professionals can not escape the fact that many of the corrective measures proposed in the past to reduce the number of overexertion and repetitive motion incidents, particularly back injuries, were not effective. Findings of the National Institute for Occupational Safety and Health (NIOSH) in the book *Work Practices for Manual Lifting* (5) have not been disputed.

Most employee selection and training procedures were questioned in the NIOSH study. The following are some examples of the NIOSH findings.

> No significant reduction in low back injuries was found by employers who used medical histories, medical examinations, or low back x-rays in selecting the worker for the job. . . .
>
> Unfortunately, no controlled epidemiological study has validated any of the contemporary theories on (lift techniques). . . .
>
> The importance of training and work experience in reducing hazard is generally accepted in the literature. The lacking ingredient is largely a definition of what the training should be and how this early experience can be given to the naive worker without harm.

There is a serious need for debate and a learned paper on manual lifting techniques, reflecting NIOSH research and other studies that put past practices in doubt. Valid guidelines for training on manual material handling would be a part of the paper.

An extensive treatise on job analysis and limiting load factors for lifting tasks is contained in the NIOSH study. Using the job analysis system proposed would logically lead to ergonomics considerations. To quote again from the NIOSH study, "More important than proper selection and training in the long-term prevention of accidents and injuries relating to lifting, is providing a safe ergonomics environment in which to work."

Hazards management recommendations now being made by safety professionals that include workplace and work methods design considerations—ergonomics—are somewhat different from those made in the past for the same types of occurrences. Since training, behavior modification, reinstruction, and adherence to safe practice rules have been the most prominent solutions recommended, safety professionals may have a credibility gap to overcome.

Ergonomics now requires attention as a specific segment of an overall safety and health program. What is the preferred order of priority in instituting an ergonomics program element? To begin with, the research necessary to identify jobs, operations, and departments that may present work station or work practice problems would be done. That implies:

1. a review of incident investigation reports and the OSHA 200 log

2. a review of workers compensation and group insurance claims experience for ergonomics related incidents

3. discussion with personnel employees concerning high-turnover jobs and jobs with excessive absenteeism

4. seeking comments from supervisors on stressful jobs

5. encouraging employees to identify jobs for which they have complaints, and

6. tours of the facility to locate jobs that:

   - require a great deal of strength or power
   - require considerable stretching, bending, or stooping
   - require considerable lifting
   - require the worker to assume awkward positions
   - are extremely repetitive
   - may present excessive vibration exposure from power tool usage
   - are performed at a rapid pace
   - present environmental discomfort (temperature, contaminants, lighting, et cetera)
   - prompt employee-generated work changes (benches, wraps on tool handles, cheater bars used on wrenches and valves, padding on chairs, footrests)
   - are monotonous
   - involve multiples of the preceding considerations

Having made those studies, hazards would be analyzed and priorities established. Consideration would be given to the most stressful jobs, and the probability and severity of injury from particular jobs. Work stations that can be easily modified and those that require capital expenditure would be identified and prioritized.

In preparing a program reflecting the findings of the studies, consideration would be given to the following in determining how ergonomics would be treated as a segment of the overall safety and health program in place.

1. Communication to management
2. Management commitment
3. Responsibility and accountability
4. Engineering involvement
5. Purchasing liaison
6. Employee involvement
7. Worksite analysis
8. Engineering modifications
9. Administrative controls
10. Personal protective equipment
11. Training

   • Engineers—first training priority
   • Supervisors
   • Employees

12. Medical management
13. Documentation, monitoring, feedback

   • Engineering revisions
   • Work practice revisions
   • Enforcement of safe work practices
   • Record keeping, analysis and review

Safety professionals should recognize that a significant change in direction is being proposed when initiating an ergonomics program.

And the concepts applicable in successfully achieving change would serve the safety professional well.

In selecting individual tasks for treatment, this excerpt from *Cumulative Trauma Disorders* gives good guidance:

There is seldom a simple, single change to be made. More often there are numerous overlapping problems involving some combination of high production demand, faulty work methods, awkward work station layouts or ill-fitting tools. . . . Perhaps the most cautious way to proceed is to administer an ergonomic intervention with the same degree of care as one would use with any new remedy. One should:

1. Perform a thorough examination first (job analysis) to determine the specific problem.

2. Evaluate and select the most appropriate intervention(s) (the assistance of an expert may be useful here).

3. Start conservative treatment (implement the intervention), on a limited scale if possible.

4. Monitor progress.

5. Continue to adjust or refine the scope of the intervention as needed. (6)

Whatever approach is taken to introduce an ergonomics program, those involved should be certain that the infrastructure is in place to respond properly to the work orders generated by awareness of ergonomics concepts. Assume that the program, having received management support, then commences directly with an employee awareness and education program. If maintenance and engineering personnel are not equipped to act on proposals for work station and work practice changes made by line employees in a reasonable time, employee enthusiasm will surely be dampened and the credibility of the ergonomics program will be damaged.

Employees must be actively involved in an ergonomics program and their participation should be considered a valuable asset. Countless success stories are being told by safety professionals about the suggestions for ergonomics workplace improvements made by line employees. Many of them are easily accomplished, inexpensive, and effective.

Only a few years ago, the conventional thinking among safety professionals was that ergonomics would not be a prominent aspect of safety practice, the assumption being that ergonomics measures were too costly in relation to benefits. History has proven that many improvements in work methods can be made without great expense. As an adjunct benefit, many companies have found that the cost to make an ergonomics change is more than offset by the increase in work efficiency.

An example of a revision in material handling methods that can be made with minimum expense is described in "Dynamic Comparison of the Two-hand Stoop and Assisted One-hand Lift Methods" (14) by Cook, Mann, and Lovested. That paper resulted from a study made at the University of Iowa to test an alternative one-hand lift method developed to address parts picking in a warehouse setting.

Learning from recent history, some cautions are offered about the introduction of ergonomics activities. First, the initial thrust should be to obtain the support of upper management, of those who make decisions concerning workplace and work methods design, and of maintenance personnel. To repeat, the infrastructure should be in place to handle what could be a flood of work orders when supervisors and line employees become participants in the ergonomics program.

Having obtained upper management awareness and support, the next course of action should be to train engineering and maintenance personnel in ergonomics fundamentals. That is the first training priority, and should be done before awareness and training programs are run for supervisors and line employees. It is folly to assume that engineering personnel, even though they may be degreed engineers, have fundamental knowledge of ergonomics principles.

Safety professionals are surely aware that OSHA is committed to the issuance of an ergonomics standard. However that gets done, when it gets done, the issuance of an OSHA standard on ergonomics could have a substantial impact on both the content of safety practice and how an ergonomics program is organized.

OSHA's target date for the issuance of a proposed ergonomics standard has been moved forward several times. Debates continue on what the standard should contain and opinions vary greatly. Comments follow on but three possibilities.

In January 1989, OSHA published voluntary, general Safety and Health Program Management Guidelines (15). While a safety professional might fault them on detail, they are generally sound. If those guidelines became a standard, an ergonomics supplement could easily be fitted to them. That represents the simplest, yet a very sound, approach.

OSHA's only official guidelines on ergonomics, the *Ergonomics Program Management Guidelines For Meatpacking Plants* (16), were issued in August 1990. The introduction recognizes the Safety and Health Program Management Guidelines issued in 1989, which "are recommended to all employers as a foundation for their safety and health programs and as a framework for their ergonomics programs."

Note that this wording implies establishing a sound safety and health program and merging ergonomics measures into it. That represents good hazards management practice.

Ergonomics Guidelines for Meatpacking duplicate much of the language in the Safety and Health Program Management Guidelines, and their structures are alike. They represent a second option. The table of contents features these elements:

I.   MANAGEMENT COMMITMENT AND EMPLOYEE
     INVOLVEMENT

    A.   Commitment by Top Management

    B.   Written Program

    C.   Employee Involvement

    D.   Regular Program Review and Evaluation

II. PROGRAM ELEMENTS

    A.   Worksite Analysis

    B.   Hazard Prevention and Control

    C.   Medical Management

    D.   Training and Education

III. DETAILED GUIDANCE AND EXAMPLES

A. Recommended Worksite Analysis Program for Ergonomics

B. Hazard Prevention and Control: Examples of Engineering Controls

C. Medical Management Program for the Prevention and Treatment of Cumulative Trauma Disorders

Ergonomics Program Management Guidelines for Meatpacking are generic. From the original table of contents, the author eliminated only two phrases. In item III. B, the phrase "For the Meat Industry" originally followed "Examples of Engineering Controls." And Item III. C had ended with the words "in Meatpacking Establishments."

In the fall of 1990, at OSHA, unpublished and unofficial *Ergonomics Program Management Recommendations For General Industry* were written. They are labeled "Internal Draft—Not for Public Release." Including attachments and a bibliography, they fill 94 double-spaced pages. About 36 of the 94 pages are devoted to the specifics of medical management. Overall, they are a bit excessive. However, they represent a third possibility: they give an indication of the thinking of some individuals at the time the recommendations were written on what should be included in OSHA's ergonomics standard.

Excerpts representing about one-eighth of the total document are included as an addendum to this essay. They duplicate a good part of the Safety and Health Program Management Guidelines. And they go into considerable detail on ergonomics methodology, in the form of a specification standard rather than a performance standard. *Nevertheless, the excerpts taken do represent the management elements of a good ergonomics program.* Preferably that program, for safety and health purposes, would be integrated into the overall hazards management program.

Legislation was proposed in 1991 to amend OSHA. It would require, within a year of enactment, the issuance of "final regulations on employer safety and health programs" and "a final standard on ergonomics hazards to protect workers from work-related musculoskeletal disorders." Whatever OSHA issues as an ergonomics standard will have a major impact on safety practice and how ergonomics programs are organized.

It should be the goal of safety professionals to be sought for their counsel by those who make workplace and work practice design decisions. Becoming well qualified in ergonomics partially prepares safety professionals for that role. Ergonomics, as it continues to emerge as having great significance on the practice of safety, has compelled many safety professionals to become students again. That's necessary to meet the challenge and to maintain a professional practice.

EXCERPTS FROM OSHA'S UNPUBLISHED AND UNOFFICIAL ERGONOMICS PROGRAM MANAGEMENT RECOMMENDATIONS FOR GENERAL INDUSTRY

EDITORIAL NOTE: All text except that in brackets is taken directly from the recommendations.

INTRODUCTION

Effective management of worker safety and health includes protection from all work-related hazards whether or not they are regulated by specific government standards.

The Occupational Safety and Health Act of 1970 (OSH Act) clearly states that the general duty of all employers is to provide their employees with a workplace free from recognized hazards.

In recent years, there has been a dramatic increase in the occurrence of . . . injuries and illnesses due to ergonomic hazards.

The incidence and severity of [ergonomic related] workplace injuries and illnesses . . . demand that effective programs be implemented to protect workers from these hazards. These [programs] should be a part of the employer's overall safety and health management program.

In January 1989 [OSHA] published voluntary, Safety and Health Program Management Guidelines, which are recommended to all employers as a foundation for their safety and health programs and as a framework for their ergonomics programs.

In addition, OSHA has developed the following general industry ergonomics program management recommendations. . . . [This program] is divided into two primary sections: a discussion of the importance of management commitment and employee involvement, followed by a recommended program with four major elements—worksite analysis, hazard prevention and control, medical management, and training and education.

The science of ergonomics seeks to adapt the job to the worker by designing tasks within the worker's capabilities and limitations.

Experience has shown that instituting programs in ergonomics has significantly reduced the incidents of [ergonomics-related] disorders and, often, improved productivity.

I. MANAGEMENT COMMITMENT AND EMPLOYEE INVOLVEMENT

Commitment and involvement are complimentary and essential elements of a sound safety and health program. Commitment by management provides the organizational resources and motivating force necessary to deal effectively with ergonomic hazards.

Employee involvement and feedback through clearly established procedures are likewise essential, both to identify existing and potential hazards and to develop and implement effective ways to abate such hazards.

A. Commitment by Top Management

The implementation of an effective ergonomics program includes a commitment by the employer to provide the visible involvement of top management.

An effective program should have a team approach with top management leading the team, and including the following:

1. Management involvement demonstrated through. . . . the priority placed on eliminating the ergonomic hazards.

2. A policy which places safety and health on the same level of importance as production.

3. Commitment to assign and communicate . . . responsibility. . . .

4. Commitment to provide adequate authority and resources. . . .

5. Commitment to ensure that [all are held] accountable for carrying out [their] responsibilities.

B. Employee Involvement

An employer should provide for and encourage employee involvement in the ergonomics program and in decisions which effect the worker safety and health, including the following:

1. An employee complaint or suggestion procedure which allows workers to bring their concerns to management and provide feedback without fear of reprisal.

2. A procedure which encourages prompt and accurate reporting of potential CTD's [and other ergonomics related injuries].

3. Safety and health committees which receive information on ergonomic problem areas. . . .

4. Ergonomic teams or monitors with the required skills. . . .

C. Written Program

Effective implementation requires a written program for job safety, health, and ergonomics that is endorsed and advocated by the highest level of management and that outlines the employer's goals and plans.

The written program should be communicated to all personnel. . . .

D. Regular Program Review and Evaluation

Procedures and mechanisms should be developed to evaluate the implementation of the ergonomics program and to monitor progress accomplished.

The results of management's reviews should be a written progress report and program update, which should be shared with all responsible parties and communicated to employees.

II. PROGRAM ELEMENTS

An effective ergonomics program should include the following four components: worksite analysis, hazard prevention and control, medical management, and training and education.

A. *Worksite Analysis*

Worksite analysis provides for both the identification of problem jobs and risk factors associated with problem jobs. The first step is to determine what jobs and work stations are the source of the greatest problems. Thus, a systematic analysis program should be initiated by reviewing injury and illness reports. . . .

The second step is to perform a more detailed analysis of those work tasks and positions previously determined to be problem areas for their own specific ergonomic risk factors. . . .

The analysis should be routinely performed by a qualified person. . . .

The analyst(s) should keep in mind the concept of multiple causation [i.e.] the combined effect of several risk factors. . . . Jobs, operations, or work stations that have multiple risk factors have a higher probability of causing [ergonomics related] disorders.

B. *Hazard Prevention and Control*

Ergonomic hazards are prevented primarily by effective design of a job or job site. An employer's program should establish procedures to correct or control ergonomic hazards using appropriate engineering, work practice, and administrative controls, coordinated and supervised by an ergonomist or a similarly qualified person.

Administrative controls reduce an employee's exposure to tasks with ergonomic hazards by schemes such as rotation to less stressful jobs, reduced production demand or quotas, and increased rest breaks.

Engineering controls, where feasible, are the preferred method of control. The primary focus of an ergonomics program is to make the job fit the person, not force the person to fit the demands of the job. This can be accomplished by redesigning the work stations, work methods, work tools, and work requirements to reduce or eliminate excessive exertion, repetitive motion, awkward postures, and other risk factors.

1. Engineering Controls

a. Principles of Work Methods

Work methods [including workstations and tools] should be designed to reduce [ergonomics] exposure. . . .

The first step is to identify the present problems [and a task analysis should follow].

b. Principles of Work Station Design

Workstations should be designed to accommodate the vast majority of the persons who work at a given job. Because workers vary considerably, it is not adequate to design for the average worker.

Workstations should be easily adjustable and designed for each specific task so that they are comfortable . . . and are appropriate for the job performed.

Specific attention should be paid to static loading of muscles, work activity height, reach requirements, force requirements, sharp or hard edges, thermal conductivity of the work surface, proper seating, support for the limbs, work piece orientation, work piece holding, and layout.

c. Tool and Handle Design

Proper attention should be paid to the selection and design of tools and workstation layouts to minimize the risk of cumulative trauma disorders (and other ergonomics related injuries).

d. Back Injury Prevention

While most back disorders result from cumulative trauma or gradual insult to the back over time, some injuries are caused by a sudden excessive load or fall. These disorders are by far the largest single category of all lost-time injuries, and have enormous financial implications.

Historically, back injury prevention has focused primarily on problems of materials handling. Common preventive measures were:

- Training workers how to lift "safely"
- Restricting the weights lifted to some maximum
- Selecting the "strongest" workers for the "heavy work"

Scientific research, industrial studies, and compensation statistics have demonstrated that these approaches have been ineffective in reducing and controlling the problem. It is now recognized that effective back injury prevention requires ongoing effort with long-term commitment to:

- Redesigning existing workplaces, jobs and equipment
- Providing training and education for all members of an organization on the causes and means of preventing back injuries as well as proper individual body mechanics.

It should be noted that there are a number of specific risk factors that may act alone or in combination to increase the risk of back disorders. . . . A list of relevant job and workstation considerations—which is by no means all inclusive—follows.

- Workplace and workstation layout
- Actions and movements
- Working posture and work positions
- Frequency and duration . . . of manual handling activities
- Load considerations

2. Work Practice Controls

An effective program for hazard prevention and control also includes procedures for safe and proper work which are understood and followed by managers, supervisors, and workers. Key elements of a good work practice program for ergonomics include proper work techniques, employee conditioning, regular monitoring, feedback, maintenance, adjustments, and modifications, and enforcement.

3. Personal Protective Equipment (PPE)

Potential ergonomic hazards should be considered when selecting PPE.

4. Administrative Controls

A sound overall ergonomics program includes administrative controls that reduce the duration, frequency, and severity of exposure to ergonomic hazards.

a. Examples of administrative controls include the following:

- Reducing the total number of repetitions
- Providing rest pauses to relieve fatigued muscles
- Increasing the number of employees assigned to the task
- Using job rotation, with caution and as a preventive measure, not in response to symptoms of cumulative trauma disorders

b. Effective programs for facility, equipment, and tool maintenance to minimize ergonomic hazards include the following measures:

- A preventive maintenance program for mechanical and power tools and equipment
- Performing maintenance regularly ...
- Effective housekeeping programs ...

C. Medical Management

GENERAL
An effective medical management program for cumulative trauma disorders is essential to the success of an employer's total ergonomic program.

The major components of a medical management program for the prevention and treatment of cumulative trauma disorders are:

1. Periodic Workplace Walkthrough

2. Symptoms Survey

3. Identification of Restricted Duty Jobs

4. Health Surveillance

5. Employee Training and Education

6. Accurate Recordkeeping

7. Periodic Program Evaluation

[EDITORIAL NOTE: In the recommendations from which these excerpts were taken, 36 pages of proposals and comments follow the preceding listing of the major components of a medical management program.]

D. Training and Education

The last major program element for an effective ergonomic program is training and education . . . to ensure that employees are sufficiently informed about the ergonomics hazards to which they may be exposed and thus are able to participate actively in their own protection.

A training program should include the following individuals:

- All affected employees
- Engineers and maintenance personnel
- Supervisors
- Managers

REFERENCES

1. Alphonse Chapanis. "To Communicate the Human Factors Message, You Have to Know What the Message Is and How To Communicate It." *Human Factors Society Bulletin,* November 1991.

2. Aetna Life and Casualty. Ergonomics Workshop. Hartford, Conn., 1990.

3. *Responding to the Workers Compensation Crisis: Can Employers Manage and Control Costs?* Valhalla, N.Y.: Towers, Perrin, Forster & Crosby, Inc., 1990.

4. *Accident Facts.* Itasca, Ill.: National Safety Council, 1991.

5. *Work Practices Guide for Manual Lifting.* Cincinnati: National Institute for Occupational Safety and Health, 1981.

6. Vern Putz-Anderson, ed. *Cumulative Trauma Disorders: A Manual for Musculoskeletal Diseases of the Upper Limbs.* Philadelphia: Taylor & Francis, 1988.

7. *Ergonomics: A Practical Guide.* Chicago: National Safety Council, 1988.

8. Suzanne H. Rodgers, technical ed. Human Factors Section, Eastman Kodak Company. *Ergonomic Design for People at Work,* vol. 1. Belmont, Calif.: Lifetime Learning Publications, 1983.

9. Suzanne H. Rodgers, principal author and technical ed. *Ergonomic Design for People at Work,* vol. 2. New York: Van Nostrand Reinhold, 1986.

10. E. Grandjean. *Fitting the Task to the Man.* Philadelphia: Taylor & Francis, 1980.

11. David C. Alexander. *The Practice and Management of Industrial Ergonomics.* Englewood Cliffs, N.J.: Prentice Hall, 1986.

12. Mustafa B. Pulat. *Fundamentals of Industrial Ergonomics.* Englewood Cliffs, N.J.: Prentice Hall, 1992.

13. Wesley E. Woodson. *Human Factors Design Handbook.* New York: McGraw-Hill, 1992.

14. T. M. Cook, S. Mann, and G. E. Lovested. "Dynamic Comparison of the Two-Hand Stoop and Assisted One-Hand Lift Methods." *Safety and Health Magazine,* National Safety Council, February 1991.

15. *Guidelines on Workplace Safety and Health Program Management.* OSHA, issued at 54 Federal Register 3904, January 26, 1989.

16. *Ergonomic Program Management Guidelines for Meatpacking Plants.* OSHA 3123, Washington, DC.: U.S. Department of Labor, 1990.

Chapter 4
Safety in the Design Process

For many years, I have believed that the greatest strides forward as respects safety, health, and the environment will be made in the design processes. And very slowly, it seems that a greater awareness has emerged of the soundness of that premise. While only a few safety professionals now give counsel on hazards in the design process, their number is increasing.

Things are happening. They suggest that safety professionals should be preparing for and, with tactful aggressiveness, be seeking greater participation in design decision making. That's where the most significant hazards related decisions are made.

A listing follows of some of the things that are happening that indicate more frequent opportunities will arise for safety professionals to give counsel in design processes.

- Some of the principles applied in quality assurance have already had an impact on the practice of safety in a few places, and that influence can be expected to be more widely felt.

- An extended adoption of the ideas on which concurrent engineering is based (also referred to as simultaneous engineering) will include—in concept and design discussions—hazard anticipation, elimination, mitigation, or control.

- OSHA's requirements for hazards analyses in "planned and new facilities, processes, materials, and equipment" will become more significant in time.

- And ergonomics, the art and science of designing work to fit the worker, will seriously effect the content of the practice of safety and require a focus on workplace and work methods design.

An organization's culture determines the probability of success of a hazards management program. And the first indications of an organization's culture in regard to hazards management are seen in the quality of its premises, production processes, products, transportation systems, environmental controls, et cetera. For some organizations, the previously listed transitions predict significant culture modifications.

Only two references will be made here to quality assurance, both coming from Mary Walton's book *The Deming Management Method* (1). First, "Quality, Dr, Deming says, must be built in at the design stage." And, "Deming's 85-15 Rule . . . holds that 85 percent of the problems in any operation are within the system and are the responsibility of management, while only 15 percent lie with the worker."

For effective hazard prevention and control, is it not also a requirement that safety must be built into systems and products in the design stage? Joe Stephenson would have you believe that to be so. In *System Safety 2000,* he stated:

> The safety of an operation is determined long before the people, procedures, and plant and hardware come together at the work site to perform a given task. . . .

> Effectively evaluating or significantly influencing the long-term success of an operation at the work site alone is virtually impossible.

> Improvements in safety can often be made for a minimal amount of money if they are made far enough upstream. Sometimes the same changes may be extremely costly to the point of being impractical or even impossible if the potential hazards or the shortcomings in the system are not recognized until the system comes together in the workplace.(2)

Deming's statement that quality must be built in at the design stage also applies, obviously, to safety. To move the practice of safety to a higher level, safety professionals must be influencing design decisions. Generally, I think Stephenson got it right. Major decisions affecting the safety of a workplace are made long before operations begin. Safety professionals should be more prominent players in those upstream decisions.

When I first entered the safety profession, almost all the work done was of an engineering nature, dealing primarily with the physical aspects of facilities and equipment. Quite often, in the literature, engineering, education, and enforcement—the three E's—were cited as the foundation of the practice of safety. And engineering was quite prominent in what we did.

Then came the behaviorists and the management systems people, who have had a significant influence on the safety profession. Their foundation has been that 88 percent of all industrial accidents were caused primarily by the unsafe acts of employees.

Some safety professionals strongly believe that safety programs can be based almost entirely on the management system idea and behavior modification, with minimum attention being given to the influence of design and engineering decisions on hazards and on incident causation.

Safety literature published in the last twenty years has contained infrequent references to the design and engineering aspects of hazards management. A few prominent writers would have you believe that behavior modification and training and management systems (consisting largely of what is referred to in OSHA literature as administrative controls) are almost the entirety of the practice of safety.

How absurd! Design and engineering considerations have largely fallen out of our concerns, unless they are imposed on us by legislation.

Yet in one subset of hazards management, the principal emphasis has been on design and engineering, at the very beginning of things. Fire protection engineers have earned recognition and are often brought into discussions by architects and engineers who seek their counsel on design specifications. Obviously, then, it can be done. Others engaged in the practice of safety can learn from the successes of fire protection professionals.

Assume for discussion that Deming's 85-15 Rule applies to safety. I believe that as a principle it does, but not with the identical numbers. Assume that safety professionals do not have an impact on the upstream side of decision making that determines the initial design of the system or on decisions effecting redesign of processes. If their work does not have a bearing on design and engineering, how much does their sphere of influence extend beyond the responsibility attributed to the worker (15 percent) into the area of management responsibility (85 percent).

Just how effective is a safety, health, and environmental program that applies only after the "system" is put in place? Such a program is principally reactive and consists mostly of data collection; compliance; inspections; safety audits; applying codes, regulations, and standards; meetings; propaganda; personal protective equipment; training; and behavior modification. Even job hazards analysis, although a highly valuable methodology, is an after-the-fact exercise.

Taking all of the typical safety and health program elements as a whole, how much of an impact can they have on the "system" for which the important safety related decisions were previously made?

Training and behavior modification have considerable prominence in safety and health program management. Are the expectations of training and behavior modification far beyond reality? We have been frequent recommenders of training as a solution for safety and health problems, even though training may not have been the appropriate remedy.

Training as a solution to problems is put in a proper perspective in *Training in the Workplace: Strategies for Improved Safety and Performance,* an excellent book by Earl D. Heath and Ted Ferry. These quotes on training are from that text. I would extend these thoughts to include behavior modification.

> Employers should not look to training as the primary method for preventing workplace incidents that result in death, injury, illness, property damage or other down grading incidents. They should see if engineering revisions can eliminate the physical safety and health hazards entirely.

So employers should understand that no amount of information and instruction, no matter how well designed and taught, can achieve a reduction in accidents unless it is designed to correct a deficiency in the skill and knowledge level of the worker. Further, employers should ensure that when defining a workplace problem to decide on appropriate remedies, they look at the design of the job, the management-supervisor-worker communications chain, and the elements of the job.(3)

Heath and Ferry recommend that engineering and design of the job be given first consideration in finding solutions to workplace problems. That is a position I strongly support.

In no way do I suggest that training does not have an important place in a hazards management program. I do propose, however, that attempts at changing the behavior of people will be of little value if the root problem is one of workplace or work methods design.

During discussions of a definition of the practice of safety, this critique was offered by Dr. Thomas A. Selders: the principal short-coming in what safety professionals do is that they seldom are in a position to anticipate hazards and give counsel on their avoidance. Dr. Selders' point was that our activity did not start soon enough in the decision stream, that it was not proactive.

So, having considered his comments and agreeing with them, I eventually wrote that the *practice of safety was accomplished through:*

- the anticipation, identification, and evaluation of hazards, and
- the giving of advice to avoid, eliminate, or control those hazards, to attain a state for which the risks are judged to be acceptable.

I emphasize "anticipation" and "giving advice to avoid". Willie Hammer spoke eloquently about the need to anticipate and to avoid or control hazards in the design process in his *Handbook of System and Product Safety:* "The system safety (and product safety) concept is predicated on this principle: The most effective means to avoid accidents during system operation is by eliminating or reducing hazards and dangers during design and development."(4)

Hammer's statement applies to every aspect of safety, whatever it's called. His premise requires the counsel of safety professionals in the design process on a proactive basis to anticipate hazards and to give advice on their avoidance, mitigation, elimination, or control.

And the counsel of safety professionals will be sought more often in the design process as the concept of concurrent engineering—also called simultaneous engineering—is more frequently adopted.

Concurrent engineering has been defined as the examination simultaneously, rather than in sequence, of the concerns of design, manufacturing, production, quality control, safety, purchasing, finance, and marketing—in the first stages of product and production process planning.

If safety professionals do not participate in design discussions, thousands of decisions that eventually determine risk levels will be made without their input. And when that occurs, the result is that safety professionals are then confronted with the workplace, processes, products, et cetera, as givens. Their work will, of necessity, be largely downstream, and reactive.

Ralph A. Evans makes these interesting comments about concurrent engineering:

> Concurrent engineering (getting all the engineers and scientists to work together, instead of separately in both time and space) is what used to be practiced; that was before bureaucracy built walls between everyone, in the name of efficiency.
>
> CE is certainly a good idea, albeit some millennia old; the only question is: How long will it be before bureaucracy gets the walls back up again? The idea that all the experts and number-crunchers should come in after a design was virtually complete, and second-guess the designers was stupid to begin with.(5)

Reading Evans, one might wonder whether concurrent engineering is simply another passing fad. The publisher of *Concurrent Engineering* by John R. Hartley would have you believe otherwise, as its "Publisher's Message" declares: "Concurrent engineering is more than the latest fad in manufacturing methodologies. It is the factor, long ignored, where true leverage for competitiveness lies."(6) Cases recited in Hartley's book indicate that there has been a widespread adoption of the concurrent engineering concept.

Of interest are the several references made to the use of Failure Modes and Effects Analysis in the design process.

In the chapter "Reliability Enhancement Measures for Design and Manufacturing" in *Concurrent Engineering and Design for Manufacture of Electronics Products* (7), Sammy G. Shina also refers to Failure Modes and Effects Analysis, and to Fault Tree Analysis.

Failure Modes and Effects Analysis and Fault Tree Analysis are System Safety tools. There is obviously a kinship here between the needs of design for manufacturability, efficiency, quality, cost minimization, maintainability—and safety.

Those few safety professionals with whom I am familiar who are involved in concurrent engineering are challenged and consider the exercise to be rewarding. In their organizations, it was decided that, as a matter of formally adopted policy, specific programs were to be established so that hazards would be addressed early on by design work groups and that safety professionals would be participants in design discussions.

Three such programs are known as Safety by Design, Process Design and Equipment Review, and Design-In Safety. In the stated purposes of those programs, designers and operations and engineering personnel are to foresee, evaluate, eliminate, and control hazards. An example of such a stated purpose follows:

> The principal thrust of the program is to develop a methodical approach for designers to foresee, evaluate and eliminate or control hazards before work on processes begins.

To serve as a model—of which there aren't very many—an example of a policy statement follows on safety in the design process. Its intent is to announce, as a matter of policy, that hazards are to be addressed during early design concept stages, and as an integral part of a concurrent engineering program.

SAFETY IN THE DESIGN PROCESS

It is our continuing policy to provide each employee with a safe work environment, and to assure a proper treatment of environmental hazards deriving from our operations.

To meet this objective, it is necessary for those personnel having design responsibilities to consider hazards, when developing new products, manufacturing processes, technology, and facilities, which may possibly impact on occupational safety and health and on the environment during the early concept stages.

It is most cost effective to design for safety, health, and environmental considerations upstream where the ability to influence is greatest. In addition to reducing risk, the concept of "Safety in the Design Process" has also been demonstrated to:

- Increase worker productivity
- Improve people and processing flexibility
- Facilitate uptime
- Reduce costs
- Reduce hazards in service and maintenance activities
- Achieve effective environmental controls, upstream

Conversely, the cost of relying on secondary engineering to retrofit for hazards impacting on safety, health, and environmental needs in the manufacturing process, after its initial design and deployment, is excessive, and often includes burdensome constraints on our manufacturing and production systems.

During the early conceptual stages inherent in product and process development, anticipating service and maintenance tasks and identifying employee exposures are critical first steps in developing the safeguarding and engineering controls necessary for protecting the employee. That same concept applies for environmental controls. It includes designing to avoid or control hazards and designing in the necessary safeguarding protection for operators and supporting maintenance personnel, and for the environment, during both planned and unplanned service of the equipment and facility.

Engineering design should strive for elimination of hazards, both in regard to the characteristics of facilities, equipment, and materials, and the possible actions or inactions of people. Only when elimination, substitution, or engineering controls are not feasible should reliance on physical barriers, warning systems, training, and personal protective equipment be considered.

The concept of "Safety in the Design Process" requires a coordinated effort between the Engineering and the Safety, Health, and Environmental communities. Our bulletin titled (xyz) establishes when safety, health, and environmental studies are necessary in the consideration of new products, technology and manufacturing processes. Please review current and future programs to assure that safety, health, and environmental issues are considered in the early stages of concept and design.

Safety, health, and environmental personnel are to assist as technical resources in achieving our "Safety in the Design Process" goals.

• • •

Application of a such a policy provides safety practitioners with opportunity for involvement, substantive accomplishment, and professional satisfaction. It places them in the design process, upstream, where the most important hazards management decisions are made.

How would an organization put into practice a policy requiring that hazards be addressed in the design process? As was the case with policy statements, there are very few implementation statements that would serve as references. This is an example that can serve as a framework for such an implementation statement.

PROCESS DESIGN AND EQUIPMENT REVIEW

PURPOSE

To provide operations, engineering, and design personnel with guidelines and methods to foresee, evaluate, and control hazards related to occupational safety and health, and the environment when considering new or redesigned equipment and process systems.

SCOPE AND DEFINITIONS

This guideline is applicable to all processes, systems, manufacturing equipment, and test fixtures regardless of size or materials used.

These conditions will be necessary for an exemption from design review:

- no hazardous materials are used (as defined by 29CFR 1910.1200);
- operating voltage of equipment is < 15 volts and the equipment will be used in nonhazardous atmospheres and dry locations;
- no hazards are present that could cause injury to personnel (e.g., overexertion, repetitive motion, error-prone situations, falls, crushing, lacerations, dismemberment, projectiles, visual injury, etc.);
- pressures in vessels or equipment are < 2 psi;
- operating temperatures do not exceed 100°F/38°C;
- no hazardous wastes as defined by 40CFR 26 & 262 and/or 331 CMR 30 are generated;
- no radioactive materials or sealed source devices are used.

If other exemptions are desired, they are to be cleared by the safety, health, and environmental professional.

PHASE I—PRE-CAPITAL REVIEW

This review is to be completed prior to submission of a project request or a request for equipment purchase, in accord with the capital levels outlined in Bulletin 246. Pre-capital reviews are crucial for planning facilities needs such as appropriateness of location, power supply, plumbing, exhaust ventilation, et cetera. Process and project feasibility are determined through this review. A complete "What If" hazard analysis, in accord with Bulletin 135, is to accompany the request. Noncapital projects should also be reviewed utilizing these procedures, but a formal "What If" hazard analysis is not required.

PHASE II—INSTALLATION REVIEW

This review requires a considerably more detailed hazards and failure analysis relative to equipment design, production systems, and operating procedures. Detailed information is documented, including equipment operating procedures, a work methods review giving emphasis to ergonomics, control systems, warning and alarm systems, et cetera. A "What If" system of hazard analysis may be used and documented. Other methods of hazards analysis will be applied if the hazards identified cannot be properly evaluated through the "What If" system.

The Project Manager shall be responsible for the establishment of a Hazard Review Committee and for managing its functions.

HAZARD REVIEW COMMITTEE

This committee will conduct all phases of design review for equipment and processes. In addition to the Project Manager, members will include the safety, health, and environmental professional, the facilities engineer, the design engineer, the manufacturing engineer, and others (financial, purchasing) as needed. For particular needs, outside consultants for equipment design or hazard analysis may be recommended by the safety, health, and environmental professional.

"WHAT IF" HAZARD ANALYSIS

This method of hazard assessment utilizes a series of questions focused on equipment, processes, materials, and operator capabilities and limitations, including possible operator failures, to determine that the system is designed to a level of acceptable risk. Users of the "What If" method would be identifying the possibility of unwanted energy release or unwanted release of hazardous materials, deriving from the characteristics of facilities, equipment, and materials and from the actions or inactions of people.

Bulletin 135 contains procedures for use of a "What If" check list. For some hazards, a "What If" check list will be inadequate and other hazard analysis methods may be used.

RESPONSIBILITIES

Project Manager
The Project Manager will be responsible for all phases of the design review, from initiation to completion. That includes initiation of the design review, forming the design review committee,

compiling and maintaining the required information, distribution of documents, setting meeting schedules and agendas, and preparation of the final design review report. Also, the Project Manager will be responsible for coordination and communication with all outside design, engineering, and hazard analysis consultants.

Department Manager

Department Managers will see that design reviews are completed for capital expenditure or equipment purchase approvals, and previous to placing equipment or processes in operation, as required under "Installation Review".

Signatures of Department Managers shall not be placed on asset documents until they are certain that all design reviews have been properly completed, and that their findings are addressed.

Design Engineer

Whether an employee or a contractor, the design engineer shall provide to the Project Manager and to the Hazard Review Committee documentation including:

- detailed equipment design drawings. . . .

- equipment installation, operation, preventive maintenance and test instructions. . . .

- details of and documentation for codes and design specifications. . . .

- requirements and information needed to establish regulatory permitting and/or registrations. . . .

For all of the foregoing, information shall clearly establish that the required consideration has been given to safety, health, and environmental matters.

Safety, Health, and Environmental Professional

Serving as a Hazard Review Committee member, the safety, health, and environmental professional will assist in identifying and evaluating hazards in the design process and provide counsel as to their avoidance, elimination, mitigation, or control. Special training programs for the review committee may be recommended by the safety, health, and environmental professional. Also, consultants may be recommended who would complete hazards analyses, other than for the "What If" system.

ADMINISTRATIVE PROCEDURES

In this section, the administrative procedures would be set forth, such as the amount of time prior to submission of a capital expenditure or equipment purchase request to be allowed the Hazard Review Committee for its work, information distribution requirements, assuring that the dates for Installation Review meetings are planned in advance, assuring that findings of hazards analyses are addressed, and resolving differences of opinion of Hazard Review Committee members.

• • •

If an organization was to adopt a policy to include safety in the design process, the preceding policy and application statements would serve well as reference sources. They are modifications, composites, and extensions of policies for safety in the design process and procedures for design review actually in place.

With an expanded impact of OSHA's requirements for hazards analyses, there will be more frequent needs and opportunities for safety professionals to become further involved in the design process. And the time is now for safety professionals to assess those possibilities, looking to the future.

OSHA's recently promulgated Standard for *Process Safety Management of Highly Hazardous Chemicals* (8) requires hazards analyses "for new facilities and for modified facilities when the modification is significant. . . ." This Standard, it's said, applies to as many as 50,000 locations.

Also, OSHA's *Safety and Health Program Management Guidelines* (9), which I believe are the framework for a standard that OSHA will promulgate, require hazards analyses of "planned and new facilities, processes, materials, and equipment."

I have written previously, in Chapter 3, on the impact that ergonomics will have on the content of the practice of safety, which will be considerable. Ergonomics compels an early involvement by safety professionals in the design process. Ergonomics, through its definition—the art and science of designing the work to fit the worker—suggests anticipation of hazards in the design process and giving counsel concerning their avoidance, elimination, mitigation, or control.

A safety professional, engaged in occupational safety and health, or in product safety, must have in-depth knowledge of ergonomics fundamentals to successfully participate in design discussions.

Whatever the source of opportunity for safety professionals to become more frequently involved in the design process, prudence requires a self-evaluation of the knowledge and skill that one would bring to the process. Being familiar with standards, codes, and regulations will serve a part of the need, but that will not be sufficient.

There are numerous checklists that will provide some help. As an indication of their breadth, these are the titles of some of them; there are many others.

- acceleration, falls and impacts
- chemical reactions
- electrical hazards
- ergonomics hazards
- explosives and explosions
- fire prevention and suppression
- lighting
- means of egress
- material handling equipment
- mechanical hazards
- mechanical lifting devices
- motor vehicles
- planning for emergencies
- pressure
- radiation hazards (ionizing and non-ionizing)
- scaffolds and ladders

- toxic materials
- ventilation
- vibration and noise
- walking and working surfaces

Checklists will, of course, be valuable. But, more important than checklists will be the concepts the safety professional has adopted concerning hazards anticipation, avoidance, elimination, mitigation, and control.

While all of the essays in this book will help in a review of those concepts, I specifically refer to the essays "On Causation Models For Hazards Related Incidents (HAZRINS)" (Chapter 11) and "Comments on Hazards" (Chapter 12) as containing some fundamental ideas.

In the essay on causation models, emphasis is given to Haddon's energy release theory (Chapters 10, 11), extended to include the release of hazardous materials. It is said that an incident causation model should give design and engineering considerations distinct and primary status, with an emphasis on avoiding unwanted releases of energy and of hazardous materials, and on ergonomics.

A section entitled "A Generic Thought Process For Hazard Avoidance, Elimination, or Control" is included in "Comments on Hazards"(Chapter 12). That outline is an adaptation of and an extension of Haddon's energy release theory. It is also recommended in that essay that safety professionals achieve an understanding of the concepts on which the MORT Safety Assurance Systems (Chapter 12) are based.

Design discussions will eventually involve determinations of the probability of hazards being realized, and what the severity of consequences will be should that occur. For preparation in design discussions, that implies having knowledge of hazard analysis and risk assessment, of risk reduction measures and their costs, of trade-offs, and of risk acceptance. Those subjects are addressed in "On Hazard Analysis and Risk Assessment"(Chapter 14).

It is proposed that all safety professionals would benefit from having an understanding of System Safety concepts, because of the successes achieved in their application and because they apply principally in the design stage. Chapter 5 is an essay titled "System Safety and the Generalist in Safety Practice".

If safety professionals do not participate in the design process, those design decisions impacting on the safety of the workplace, on work methods, on environmental controls, on products, and the like will continue to be made without their input. Activity of safety professionals will remain on the "downstream" side of the important decisions.

REFERENCES

1. Mary Walton. *The Deming Management Method.* New York: Putnam Publishing Group, 1986.

2. Joe Stephenson. *System Safety 2000.* New York: Van Nostrand Reinhold, 1991.

3. Earl D. Heath and Ted Ferry. *Training in the Work Place: Strategies for Improved Safety and Performance.* Goshen, N.Y.: Aloray, 1990.

4. Willie Hammer. *Handbook of System and Product Safety.* Englewood Cliffs, N.J.: Prentice-Hall, 1972.

5. R. A. Evans. *IEEE Transactions on Reliability,* a publication of the IEEE Reliability Society, March 1992.

6. John R. Hartley. *Concurrent Engineering.* Cambridge, Mass.: Productivity Press, 1992.

7. Sammy G. Shina. *Concurrent Engineering and Design for Manufacture of Electronic Products.* New York: Van Nostrand Reinhold, 1991.

8. *Process Safety Management of Highly Hazardous Chemicals.* OSHA Standard, 1910.119, February 1992.

9. *Safety and Health Program Management Guidelines.* OSHA, January 1992.

10. William J. Haddon, Jr. *The Prevention of Accidents: Preventive Medicine.* Boston: Little, Brown, 1966.

11. William J. Haddon, Jr. "On the Escape of Tigers: An Ecological Note." *Technology Review*, May 1970.

12. William G. Johnson. *MORT Safety Assurance Systems*. New York: Marcel Dekker, 1980.

Chapter 5

System Safety and the Generalist in Safety Practice

Although they have been successful, those system safety professionals, very few of us engaged in the general practice of safety have adopted their concepts and techniques. I do not suggest that the generalist in safety practice must become a specialist in system safety. But as a matter of personal professional development, I suggest that we generalists would improve the quality of our performance by acquiring knowledge of what system safety is all about.

In *MORT Safety Assurance Systems,* William G. Johnson expressed an opinion (with which I agree) about accomplishments that could not have been achieved without applying system safety concepts: "The system safety programs used in aerospace, nuclear, and military projects provided a well-ordered guide to some requirements for a superlative effort. Indeed, they are a route to accomplishing things which would otherwise be beyond human reach." (1)

Those accomplishments are a matter of fact. And I believe that they are immense. Anyone who has an understanding of risk, the complexity of the hardware involved, and the demands on personnel must marvel at the success of a space shot.

What can we learn about the system safety concepts and methodologies applied in those achievements that would be of benefit in the general practice of safety?

At least one other author expected a more widespread adoption of system safety concepts, beyond the use by the military, aerospace personnel, and nuclear facility designers, and had to recognize that it wasn't happening. In *The Loss Rate Concept In Safety Engineering*, R. L. Browning wrote:

> As every loss event results from the interactions of elements in a system, it follows that all safety is "systems safety." . . . The safety community instinctively welcomed the systems concept when it appeared during the stagnating performance of the mid-1960s, as evidenced by the ensuing freshet of symposia and literature. For a time, it was thought that this seemingly novel approach could reestablish the continuing improvement that the public had become accustomed to; however, this anticipation has not been fulfilled.

> Now, almost two decades later, although systems techniques continue to find application and development in exotic programs (missiles, aerospace, nuclear power) and in the academic community, they are seldom if ever met in the domain of traditional industrial and general safety. (2)

Although there were countless seminars and a proliferation of papers on system safety, the generalist safety professional seldom adopted system safety concepts.

In response to his own question—Why this rejection?—Browning said that traditional safety is predicated on absolutes, "safe" or "unsafe," while the concepts of measurable and acceptable risk are fundamental to system safety. Also, he expressed the view that system safety literature and training may have turned off generalist safety professionals because of the "exotica" they usually presented.

In *MORT Safety Assurance Systems* (1), Johnson cited two obstacles that deter the adoption of system safety concepts in industrial safety practice: the noncontinuous nature of system safety work, being project-oriented; and system safety methods applied to other than significant hazards can represent "overkill."

Nevertheless, Browning went on to build *The Loss Rate Concept of Safety Engineering* (2) on system safety concepts. He also gave this encouragement:

Furthermore, we have found through practical experience that industrial and general safety can be engineered, at a level considerably below that required by the exotics, using the mathematical capabilities possessed by average technically minded persons, together with readily available input data.

And system safety concepts were significant in the development of *MORT*. References to those concepts will be found throughout Johnson's text.

There is much reality in the Browning and Johnson observations. System safety literature is loaded with governmental jargon and can easily repel the uninitiated. It makes more of individual system safety methodologies, especially of highly complex methods requiring extensive knowledge of mathematics and the use of probability estimates, than it does of concepts and purposes. Some of the system safety literature does give the appearance of exotica. And the use of some of the methodologies for analyses of hazards of lesser significance can be very expensive.

System safety professionals, in their literature, have not done a good job of marketing their concepts, or of defining system safety. Nor have they agreed on a definition of their practice.

With the hope of generating a further interest by generalist safety professionals in system safety, I shall emphasize the concepts and purposes through which successes have been achieved rather than explore methodologies in detail. As Ted Ferry wrote in his preface to Joe Stephenson's *System Safety 2000*, "Professional credentials or experience in 'system safety' is not required to appreciate the potential value of the systems approach and system safety techniques to general safety and health practice." (3) To paraphrase Browning, all hazards-related incidents result from interactions of elements in a system. Therefore, all safety is system safety. Therein lies an important idea.

Others have suggested that the system idea is what is important and that system safety purposes could be met, in many instances, with knowledge of sound safety practice and with intuition.

In *Safety Management,* John V. Grimaldi and Rollin H. Simonds wrote: "System safety analyses require the imaginative construction of every conceivable situation that could arise. . . . A reference to system analysis may merely imply an orderly examination of an established system or subsystem." (4)

In "Ergonomics Aids Industrial Accident and Injury Control," Richard G. Pearson and Mahmoud A. Ayoub gave this view:

> By systems approach, we mean a conscientious, systematic effort to design an effective system, such as a production plant, giving due consideration to the interaction among man, machine, and environment. From the ergonomics viewpoint, prime considera-tion is given to human performance and safety considerations. A cardinal principle of ergonomics is that since everything is designed ultimately for man's use or consumption, man's charac-teristics should be considered from the very beginning of the design cycle. . . .
>
> As inferred earlier under "The systems approach," systems evalu-ation should be a continuous process during the design, develop-ment, installation, operation, and maintenance of an industrial manufacturing system. (5)

How would I define the system safety idea, considering the ref-erences cited and other essays in this book? Consider this.

THE SYSTEM SAFETY IDEA

1. Safety is a state for which the risks are judged to be acceptable.
2. All risks, in the context of system safety, derive from hazards.
3. Hazards are the justification for existence of the entirety of safety practice: if there are no hazards, safety professionals need not exist.
4. Unwanted releases of energy and unwanted releases of haz-ardous materials are the causal factors for hazards-related inci-dents.
5. All hazards-related incidents result from interaction of ele-ments in a *system*.
6. Hazards are most effectively and economically anticipated, avoided, mitigated, or controlled in the initial system design process, or in the redesign of existing systems.

7. To design for minimum risk and to achieve a state in which the risks are judged to be acceptable, the order of effectiveness, *in developing or modifying a product or system,* is to

- Design for minimum risk: avoid, eliminate, mitigate, or control hazards in the design or redesign processes, giving due consideration to ergonomics
- Incorporate designed and engineered safety devices
- Provide warning devices
- Develop operating procedures, with a particular recognition of the capabilities and limitations of personnel
- Provide for knowledge acquisition and training

8. An orderly and continuous hazards management process is to apply *throughout* the design, development, installation, operation, maintenance, and disposal of a product or system.

9. Safety professionals exist for these purposes only:

- to anticipate, identify, and evaluate hazards, and
- to give advice on the avoidance, elimination, or control of hazards, to attain a state for which the risks are judged to be acceptable

10. To achieve their purposes, generalists in safety practice would adopt a *system safety concept* that includes

- An understanding of hazards
- An understanding of risk
- Addressing hazards and risks within their *systems*
- An understanding of unwanted releases of energy and unwanted releases of hazardous materials being the causal factors for hazards related incidents
- Knowledge, through which optimum safety can be attained in the design or redesign of systems or products, considering the life cycle of a product or system, in
  - applied engineering
  - applied science
  - applied management

- legal and regulatory requirements and other design criteria
- hazards analyses and risk assessment methods

I'm aware that my outline of *The System Safety Idea* does not fit precisely with any of the several definitions of system safety published. It encompasses most, and goes beyond several. I hope it's of interest to more generalists in the practice of safety than are now applying system safety concepts. I am certain that application of system safety concepts in the industrial setting would result in significant reductions in injury and illness incidents.

System safety is hazards based. So are all subsets of the practice of safety, whatever they are called. System safety commences with hazard identification and analysis. Do that poorly, and all that follows is misdirected. Hazard analysis methods used in system safety have been successful, and the generalist in safety practice ought to know more about them.

System safety concepts apply in the design and redesign processes. I propose that having knowledge of system safety concepts will better prepare the generalist safety professional to successfully give counsel during design endeavors.

Ernest Levins, during his tenure as Director of Safety at McDonald-Douglas in Santa Monica, gave his views on an effective hazard-analysis scheme in an article titled "Search I—Fourth Installment, Locating Hazards Before They Become Accidents." His outline follows.

SEVEN STEPS TO FOLLOW

The following scheme for an effective hazard-analysis embodies the concepts discussed here:

1. Define the system in space and time—including its objectives, the location of its interfaces with other systems of interest, and the analytical limits of resolution within the system (may vary, depending on the analyst's interests).

2. Specify the identifiable undesired outcomes, states, or conditions within the defined system.

3. Select the key undesired system-characteristic outcomes that serve as the basis for decision—by comparing them with some index of criticality (e.g., negligible, marginal, critical, catastrophic); record the results.

4. Determine the possible modes of occurrence of the selected undesired outcomes.

5. Evaluate the likelihood of occurrence of the possible modes of occurrence. This can be done with logic alone; failure rate data are not needed in most industrial systems. An estimate of the mission-success can even be made on the basis of events being "likely" or "unlikely" to occur.

6. From the foregoing, decide if the system design is adequate to prevent failure; and, if not, what design changes are required to improve relative safety of the system.

7. Analyze any system revision as above, and repeat as often as necessary until the optimum design is achieved. (6)

Levins speaks of "relative safety" and "optimum design." What do those terms imply? Design goals are not intended to attain a risk free system, which is unattainable. And judgments will be made concerning the consequence and probability of hazard realization, costs of risk reduction, performance requirements, and scheduling in arriving at the optimum design.

Safety professionals must realize that resources required to allow identifying and evaluating, and eliminating or controlling every hazard will never exist.

Levins does suggest that some of the judgments necessary can be made with logic alone. That's important. System safety concepts can be implemented in many cases without applying detailed and extensive hazard analysis methods. But logic alone will not be sufficient for all hazards. For some hazards, applying the hazard analysis and risk assessment methods specifically developed as a part of system safety will be necessary.

A bit of history on the evolution of system safety will give an insight into its origins, the need for the hazard analysis and risk assessment techniques developed, and the place that system safety has attained. Authors don't agree on when or where it all started.

But all the historical references on system safety do relate to the military or to aeronautics.

There are many citings in the literature of adverse accident experience in the military branches, which are said to have given impetus to the development of system safety concepts. This example is taken from "Why 'System Safety'" by Charles O. Miller, Director in 1971 of the Bureau of Aviation Safety at the National Transportation Safety Board:

> Statistics show that far more aircraft, and indeed more pilots, were lost in stateside operations during [World War II] than ever were in combat. In 1943, for example, something like 5,000 aircraft were destroyed stateside, against 3,800 in the war proper. . . .
>
> Then, shortly after World War II, in the period 1946 to 1948, people in the military were astonished by a new accident peak. . . .
>
> Thus, the war experience, plus the immediate postwar experience, resulted in a call to the technical community for help. (7)

But, in his *Handbook of System and Product Safety*, Willie Hammer states: "Oddly enough, it was more the concern with unmanned systems, the intercontinental ballistic missiles (ICBMs), that led to the development of the system safety concept."(8)

C. W. Childs in "Industrial Accident Prevention Through System Safety" gives this brief history of the origins of system safety.

> In the early stages of missile development, it was necessary to assemble thousands of subsystems and millions of component parts in such a manner as to be virtually free or at least "forgiving" of mistakes or failures. The management and engineering disciplines at this point in time were not sophisticated or rigid enough to assure such a concept.
>
> Consequently, there were some very spectacular accidents and the engineering disciplines of reliability and quality control were tasked to provide a greater measure of mission success through elaborate quality assurance and reliability analyses and testing. Despite this, accidents continued at a higher rate than was considered acceptable and the new discipline of system safety was

developed to bring the accident preventive experience into the systems engineering processes.

Since that time, some methods and techniques have been developed which not only resulted in decreased accident rates for complex flight systems, but are now being used to prevent accidents in the industrial environment." (9)

And great successes were achieved. But I do wish that Childs' latter statement—that system safety concepts applied in developing complex flight systems were being used to prevent accidents in the industrial environment—could be substantiated more than I have been able to do. Childs' paper was published in 1974. Nevertheless, he gives a valuable historical perspective.

There were many system safety developments in the military and at the National Aeronautical and Space Administration (NASA) in the 1960s, '70s and '80s. Several military standards were issued on system safety engineering or safety engineering of systems, the most prominent being MIL-STD-882B. Its full title is Military Standard—System Safety Program Requirements (10). Its most recent modification was made in 1987.

You must have a copy of MIL-STD-882B. I believe it to be an exceptionally valuable reference. I shall extract from it with the hope that generalist safety professionals will agree that its premises are applicable in their fields. It is a rarity for a significant publication on system safety, or a chapter on system safety in a book, to be written without being largely based on the requirements of this standard.

MIL-STD-882B is approved for use by all departments and agencies of the Department of Defense. Many organizations within NASA also apply the standard in principle. Each contractor to which it applies, providing services or equipment, is to have a System Safety Program Plan that describes the methods to be followed, tailored to the contractor's operations, in accomplishing the requirements of a System Safety Program.

Objectives of the System Safety Program are specifically outlined. *Those same objectives apply in generalist safety practice.* They are restated here, with modifications to eliminate some of the governmental jargon.

System Safety Program Objectives

The system safety program shall define a systematic approach to make sure:

a. Safety, consistent with objectives, is designed into the system in a timely, cost-effective manner.

b. Hazards associated with each system are identified, evaluated, and eliminated, or the associated risk reduced to an acceptable level.

c. Historical safety data, including lessons learned from other systems, are considered and used.

d. Minimum risk is sought in accepting and using new designs, materials, and production and test techniques.

e. Actions taken to eliminate hazards or reduce risk to an acceptable level are documented.

f. Retrofit actions required to improve safety are minimized through the timely inclusion of safety features during research and development, and in the purchasing and acquisition process.

g. Changes in design, configuration, or objectives are accomplished in a manner that maintains an acceptable level of risk.

h. Consideration is given to safety, and the ease and methods of disposal, of any hazardous materials.

i. Significant safety data representing lessons learned through program activity are documented and retained for future reference, and disseminated to others.

Then MIL-STD-882B sets forth system safety design requirements that are to be specified after a review of pertinent standards, specifications, regulations, design handbooks, and other sources of guidance applicable to the design of the system. This is a slight modification of those requirements.

System Safety Design Requirements

Some general system safety design requirements are:

a. Eliminate identified hazards or reduce associated risk through design, including material selection or substitution. When hazardous materials must be used, select those with the least risk throughout the life cycle of the system.

b. Isolate hazardous substances, components, and operations from other activities, areas, personnel, and incompatible materials.

c. Locate equipment so that access during operations, servicing, maintenance, repair, or adjustment minimizes personnel exposure to hazards (e.g., hazardous chemicals, high voltage, electromagnetic radiation, cutting edges, or sharp points).

d. Minimize risk resulting from excessive environmental conditions (e.g., temperature, pressure, noise, toxicity, acceleration, and vibration).

e. Design to minimize risk created by human error in the operation and support of the system.

f. Consider alternate approaches to minimize risk from hazards that cannot be eliminated. Such approaches include interlocks, redundancy, fail-safe design, fire suppression, and protective clothing, equipment, devices, and procedures.

g. Protect the power sources, controls and critical components of redundant systems by physical separation or shielding.

h. To ensure personnel and equipment protection when alternate design approaches cannot eliminate the hazard, provide warning and caution notes in assembly, operations, maintenance, and repair instructions, and distinctive markings on hazardous components and materials, equipment, and facilities. Throughout, warning and caution notes and markings shall be standardized.

i. Minimize the severity of personnel injury or damage to equipment, should a hazards-related incident occur.

j. Design software controlled or monitored functions to minimize initiation of hazards-related incidents.

k. Review design criteria for inadequate or overly restrictive requirements regarding safety and recommend new design criteria, supported by study and analyses.

These design requirements could serve well, in my opinion, in any industrial setting. They are compatible with "A Generic

Thought Process for Hazard Avoidance, Elimination, or Control," which is included in "Comments on Hazards," (Chapter 12). That outline is an extension of Haddon's energy release theory.

Next in the standard comes an order of precedence in satisfying system safety requirements. Its principle points are contained, and extended, in what I referred to as the *order of effectiveness* that previously appeared in this essay. For both the order of precedence and the order of effectiveness, the theme is to design for minimum risk.

Risk assessment is the next major caption in the system safety requirements. There is but one paragraph, best summarized as follows:

Risk Assessment

Decisions regarding resolution of identified hazards shall be based on assessment of the risk involved. Hazards shall be characterized as to the worst credible hazard severity and hazard probability. The priority for system safety is eliminating hazards by design. In the early design phase, a risk assessment considering only hazard severity will generally suffice. When hazards are not eliminated during the early design phase, a risk assessment shall be made that addresses both hazard severity and hazard probability in establishing methods for resolution of identified hazards.

The hazard severity categories given are catastrophic, critical, marginal, and negligible. Hazard probability categories, for a unit of time or activity, are frequent, probable, occasional, remote, and improbable. In "On Hazard Analysis and Risk Assessment," (Chapter 14), I replaced the term hazard severity with severity of consequence and hazard probability was replaced with occurrence probability. MIL-STD-882B was the source for severity and probability categories, as was the matrix in that essay.

It is required by the standard that action be taken to eliminate identified hazards or reduce their associated risks. Those risks categorized as catastrophic and critical are to be eliminated, or their associated risks reduced to an acceptable level. If that is impossible or impractical, alternatives are to be recommended.

Three task sections follow in the standard, "Program Management and Control," "Design and Evaluation", and "Software Hazard Analysis."

References on the requirements for applying specific hazards analyses techniques are found in "Design and Evaluation." Some of the techniques mentioned are Preliminary Hazard List, Preliminary Hazard Analysis, System Hazard Analysis, Operating Hazard Analysis, Occupational Health Hazard Assessment, and Top-Level Design Hazard Analysis.

This Military Standard for System Safety Program Requirements is quite a document. I again recommend it to generalist safety professionals as a resource. It is design and engineering oriented; in that respect, its general premises also apply to the general practice of safety. Understanding and adopting those general premises is what I propose. In doing so, the generalist would acquire basic knowledge of the formal hazard analysis techniques that have been developed in system safety and when they should be recommended. Such analyses would be completed by system safety specialists.

It would be unprofessional to present oneself as having a capability not possessed. It would be more unprofessional to attempt an exercise for which one was unqualified, and produce an inadequate product which was the basis of management decisions.

In *Managing Risk: Systematic Loss Prevention For Executives,* Vernon L. Grose expressed a sound caution that applies to all endeavors involving hazards analyses:

> Preparing an HMEA [Hazard Mode and Effects Analysis] is a complex process. . . . An important caution (and one frequently violated); an HMEA should never be prepared by risk management personnel who possess only general knowledge. If they do— instead of seeing that it is done—the analysis will never have credibility with the very people who must take preventive action. On the other hand, risk management's contribution is to
>
> (1)   alert executive management when such an analysis is required,
>
> (2)   provide the analytical methodology,

(3)   oversee the analysis,

(4)   assure the analysis is carried through to its intended goal. . . . (11)

What Grose has written as being applicable to "risk management personnel" defines precisely what I believe should be an operational outline for safety generalists. They should not undertake to do what they are not qualified to do. They should be able to determine when an in-depth risk analysis is required and when specialized talent should be engaged for that purpose. To fulfill that role, knowledge is required of the various analytical systems and their advantages and shortcomings.

According to Grose, the success of Project Apollo (the manned exploration of the moon) is partially credited to HMEA since it was the primary type of analysis employed to foresee hazards and their associated risks. He refers to HMEA as the granddaddy of in-depth risk analysis. Failure Modes and Effects Analysis (FMEA), Grose says, is one of its several names. FMEA is one of the analytical methods with which the safety generalist should be knowledgeable.

For safety generalists who take an interest in system safety concepts, I offer this short list of readings:

1. As a primer, MIL-STD-882B (10). Purposes and task descriptions for analytical techniques are included.

2. Also, as a primer, an article by Pat Clemens titled "A Compendium of Hazard Identification & Evaluation Techniques for System Safety Application" (12), which comments on twenty-five analytical methods.

3. *Handbook of System and Product Safety* (8), by Willie Hammer. This is one of the first texts on system safety and its methods.

4. *System Safety Engineering and Management* (13), by Harold E. Roland and Brian Moriarty, gives an overview of a system safety program and describes several analytical techniques, including Fault Tree Analysis. Contrary to what some others have said, Fault Tree Analysis is not a branch of metaphysics.

5. *The Loss Rate Concept In Safety Engineering* (2), by R. L. Browning. This is a good little book to which I have referred sev-

eral times. Browning believes that one can apply system safety concepts in an industrial setting without necessarily becoming exotic. He builds on the Energy Cause Concept and works through qualitative and quantitative analytical systems, including Fault Tree Analysis.

6. *Managing Risk: Systematic Loss Prevention for Executives* (11), by Vernon L. Grose. Grose takes a system approach for hazard identification, the writing of scenarios concerning them, and judgments made by "juries" of qualified personnel that consider the scenarios as to probability of occurrence, severity of outcome, and the cost of risk control. Rankings, which are non-numerically quantified, are given to risks according to a Hazard Totem Pole. Grose is leery of the "numerologists".

7. *System Safety 2000* (3), by Joe Stephenson. Beginning with a history and the fundamentals of system safety, the author then moves into system safety program planning and management, and system safety analysis techniques. About half of the book is devoted to those techniques. A safety generalist would find it to be a good and not too difficult read.

It is my purpose to:

- recognize the many successes that have been achieved through the application of system safety concepts
- establish that fundamental system safety concepts can be applied by generalists in safety practice
- outline *The System Safety Idea*
- review system safety program objectives and system safety design requirements and relate them to the general practice of safety
- encourage generalists who have not adopted system safety concepts in their practices to commence inquiry and education to do so

I sincerely believe that we generalists in safety practice can learn from system safety successes and be more effective in our work through adopting system safety concepts.

REFERENCES

1. William G. Johnson. *MORT Safety Assurance Systems*. New York: Marcel Dekker, 1980.

2. R. L. Browning. *The Loss Rate Concept in Safety Engineering*. New York: Marcel Dekker, 1980.

3. Joe Stephenson. *System Safety 2000*. New York: Van Nostrand Reinhold, 1991.

4. John V. Grimaldi and Rollin H. Simonds. *Safety Management*. Homewood, Ill.: Irwin, 1989.

5. Richard G. Pearson and Mahmoud A. Ayoub. "Ergonomics Aids Industrial Accident and Injury Control." *IE*, June 1975.

6. Ernest Levins. "Search I—Fourth Installment: Locating Hazards Before They Become Accidents." *Journal of the American Society of Safety Engineers*, May 1970.

7. Charles O. Miller. "Why 'System Safety.'" *Technology Review*, February 1971.

8. Willie Hammer. *Handbook of System and Product Safety*. Englewood Cliffs, N.J.: Prentice-Hall, 1972.

9. C. W. Childs. "Industrial Accident Prevention Through System Safety." *Hazard Prevention*, September—October 1974.

10. *Military Standard—System Safety Program Requirements*, MIL-STD-882B. Department of Defense, Washington, D.C., 1987.

11. Vernon L. Grose. *Managing Risk: Systematic Loss Prevention For Executives*. Englewood Cliffs, N.J.: Prentice-Hall, 1972.

14. P. L. Clemens. "A Compendium of Hazard Identification and Evaluation Techniques for System Safety Applications." *Hazard Prevention*, March—April 1982.

15. Harold E. Roland and Brian Moriarty. *System Safety Engineering and Management*. New York: John Wiley & Sons, 1983.

Chapter 6
On the Practice of Safety and Total Quality Management

"Safety Professionals: Things Are Happening" (Chapter 1) was written reflecting comments made by safety professionals in response to inquiries seeking their views on major changes that could have an impact on the practice of safety. Some safety professionals mentioned having been recently introduced to and given assignments in total quality management and that it should be expected that other safety professionals will become similarly involved. Total quality management has a remarkable kinship with hazards management.

Surely, American industry has been on a drive to improve quality. The Malcolm Baldrige National Quality Award (1) given by the United States Department of Commerce, is a mark of great accomplishment and prestige. Competition for the award is keen: for 1991, over 240,000 copies of the criteria for the award were requested of the Department of Commerce and it's expected that the number for 1992 will be close to that for 1991.

Those safety professionals who have been brought into quality management programs express pleasure and satisfaction with their involvement. Everyone one of them with whom I have spoken says that the management practices applied in quality management are almost identical with those required to achieve effective hazards management. My review of the literature confirms that opinion.

I cannot say that a large number of safety professionals now participate in total quality management teams. I do say that it is one of the processes through which things get done for which safety professionals should seek participation.

Involvement in total quality management presents considerable opportunity for accomplishment and improved effectiveness through participation in solving problems that broadly impact on productivity, quality, and safety, and for greater recognition by management.

Just what is Total Quality Management? How did I conclude that its management practices have a remarkable kinship to hazards management? In discussions with safety professionals presently involved with quality management, mention was frequently made of the program through which the Malcolm Baldrige National Quality Award is given; the work of two authors, W. Edwards Deming and Philip B. Crosby; and the Six Sigma quality management program at Motorola, the 1988 winner of the Baldrige Award.

This essay is a summary resulting from a review of those programs, texts written by the two authors cited, and of other related texts.

As the following outline for a quality management program is reviewed, the requisites of a hazards management program should be kept in mind. This outline appears in the 1992 Criteria for the Malcolm Baldrige National Quality Award issued by the U.S. Department of Commerce. Its program is administered by the American Society for Quality Control.

1992 Examination Categories/Items

1.0 Leadership
- 1.1 Senior Executive Leadership
- 1.2 Management for Quality
- 1.3 Public Responsibility

2.0 Information and Analysis
- 2.1 Scope and Management of Quality and Performance Data and Information

This seems an ideal beginning outline for a hazards management program. Whether or not an organization is seeking the Baldrige

Award, if the purposes of its quality management program are similar to those set forth in this outline, there is much opportunity for involvement by safety professionals. In many respects, the management processes in which the safety professional should have an interest are identical with those that are of interest to personnel responsible for quality. It seems that all personnel are involved in those companies that have undertaken to improve quality. I hear of "teams". Safety professionals could accomplish much as members of those "teams."

Although the term customer does not appear in our literature, safety professionals have to consider two categories of customers— customers within operations, and customers external to operations who might be effected by them.

Each of the seven categories for the Baldrige Award is allocated a number of points, the possible total being 1000. Below is an excerpt from the scoring system for a few of the categories. In substance they represent ideals for safety, health, and environmental management programs.

1.1 Senior Executive Leadership
Describe the senior executives' leadership, personal involvement, and visibility in developing and maintaining a customer focus and an environment for quality excellence.

1.2 Management for Quality
Describe how the company's customer focus and quality values are integrated into day-to-day leadership, management, and supervision of all company units.

1.3 Public Responsibility
Describe how the company includes its responsibilities to the public for health, safety, environmental protection, and ethical business practices in its quality policies and improvement activities, and how it provides leadership in external groups.

4.2 Employee Involvement
Describe the means available for all employees to contribute effectively to meeting the company's quality and performance objectives; summarize trends in involvement.

5.1 Design and Introduction of Quality Products and Services
Describe how new and/or improved products and services are designed and introduced and how processes are designed to meet key product and service requirements and company performance requirements.

5.5 Quality Assessment
Describe how the company assesses the quality and performance of its systems, processes, and practices and the quality of its products and services.

If what is expected in an organization with respect to quality management resembles the Baldrige Award requirements, it would seem prudent for safety professionals to be extensively involved. It's of interest that there has been some talk in Washington about developing a similar award program for safety.

Although W. Edwards Deming, author of *Out of the Crisis* (2), is considered by many an expert in quality assurance, it occurred to me as I read his book that he might view such a designation as rather narrow in relation to what his work is intended to do.

Many authors have quoted Deming's "14 points," not necessarily with precision. In *Out of the Crisis,* Deming gives a "Condensation of the 14 Points of Management," beginning with this statement: "The 14 points are the basis for transformation of American industry." Deming outlines a management theory he believes must be applied if American industry is to be successful in the world market. Within that theory is his framework for attaining superior quality. The following excerpt presents Deming's 14 points directly and entirely.

Condensation of the 14 Points of Management

1. Create constancy of purpose toward improvement of product and service, with the aim to become competitive and to stay in business, and to provide jobs.

2. Adopt the new philosophy. We are in a new economic age. Western management must awaken to the challenge, must learn their responsibilities, and take on leadership for change.

3. Cease dependence on inspection to achieve quality. Eliminate the need for inspection on a mass basis by building quality into the product in the first place.

4. End the practice of awarding business on the basis of price tag. Instead, minimize total cost. Move toward a single supplier for any one item, on a long-term relationship of loyalty and trust.

5. Improve constantly and forever the system of production and service, to improve quality and productivity, and thus constantly decrease costs.

6. Institute training on the job.

7. Institute leadership. The aim of supervision should be to help people and machines and gadgets to do a better job. Supervision of management is in need of overhaul, as well as supervision of production workers.

8. Drive out fear, so that everyone may work effectively for the company.

9. Break down barriers between departments. People in research, design, sales, and production must work as a team, to foresee problems of production and in use that may be encountered with the product or service.

10. Eliminate slogans, exhortations, and targets for the work force asking for zero defects and new levels of productivity. Such exhortations only create adversarial relationships, as the bulk of the causes of low quality and low productivity belong to the system and thus lie beyond the power of the work force.

11a. Eliminate work standards (quotas) on the factory floor. Substitute leadership.

b. Eliminate management by objective. Eliminate management by numbers, numerical goals. Substitute leadership.

12a. Remove barriers that rob the hourly worker of his right to pride of workmanship. The responsibility of supervisors must be changed from sheer numbers to quality.

b. Remove barriers that rob people in management and in engineering of their right to pride of workmanship. This means, inter alia, abolishment of the annual or merit rating and of management by objective.

13. Institute a vigorous program of education and self-improvement.

14. Put everybody in the company to work to accomplish the transformation. The transformation is everybody's job.

In *Out of the Crisis,* Deming expounds in detail on his management theory. Some companies have built their quality improvement programs on much of what Deming has proposed. But I do not know of an organization that has adopted all of Deming's 14 points.

As Rafael Aguayo wrote in *Dr. Deming—The American Who Taught The Japanese About Quality* (3), "The management lessons of Deming are in direct opposition to what is currently taught in most business schools and advocated by management consultants and business writers." (3) Deming would do away with all performance reviews, in any form, and performance incentives. In two places in his 14 points, he records his opposition to management by objectives.

Nevertheless, a great deal of what Deming outlines as management theory is directly related to the principles and practices that would be applied in attaining successful hazards management. After reviewing *Out of the Crisis, Dr. Deming—The American Who Taught The Japanese About Quality,* and a book by Mary Walton titled *The Deming Management Method* (4), I compiled the following summary of the particularly relevant points in those writings that relate closely to hazards management. Wherever the word *quality* appears in this summary, *safety* can be substituted.

- Quality begins in the board room.
- Significant improvement in quality requires a change in the corporate culture.
- A long term commitment by management and knowledge of what actions must be taken are necessary to measurably improve quality.

- Management support, alone, will not be sufficient. Personal management action and leadership are required. Management obligations can not be delegated.

- Only management can initiate improvement in quality and productivity. Workers are helpless to change the systems in which they work. On their own, they can achieve very little.

- Everybody has customers. If a person is not aware of whom the customers are, that person does not understand the job.

- Quality must be built in at the design stage, where teamwork is fundamental. When processes are in place, revisions are costly. Quality is achieved from a continuing improvement in processes.

- Quality comes not from inspection but from improvement of the process. The old way: Inspect bad quality out. The new way: Build good quality in.

- Inspection to improve quality is performed too late in the system. Mass inspection is ineffective and costly. (There are exceptions to this dictum).

- Fundamental changes in the system can significantly reduce variations, not adjustments in an existing system.

- Managers have a tendency to put responsibility on workers that is beyond their control. People work in the system that is created by management and that only management can change.

- Employees are not a part of the decision making for such considerations as plant layout, lighting, heating, ventilation, product design, process selection and design, determination of work methods, and equipment and materials purchasing. Why should they be held responsible for quality?

- Decisions to improve a system should relate to statistical knowledge and thinking. Basing decisions on timely and accurate data is critical. Intelligent decisions can be made only when the data is accurate and properly interpreted. Relying only on data, though, can lead to difficulty.

- Meeting established specifications ensures maintaining the present status. It does not result in improvement.

- Everyone in management must be educated by a teacher, qualified academically and in work experience, in statistical techniques. Everyone includes "all engineers and scientists, inspectors, quality control managers, accounting, payroll, purchase, safety, legal department, consumer service, consumer research."

- Understanding the distinction between a stable system and an unstable system is essential. A statistical chart, with points properly charted, indicates whether or not a system is stable.

- Common causes are faults of the system (capabilities of the equipment, work methods design, unclear procedures), and often can only be corrected by management. Special causes (a machine malfunctions, an untrained worker is assigned) derive from fleeting events. As to troubles and improvement possibilities, it is estimated that 94 percent are common causes and in the system (responsibility of management), and 6 percent derive from special causes. Special causes should be sought and corrected at once.

- Quoted from *Out of the Crisis:*

  The first step in reduction of the frequency of accidents is to determine whether the cause of an accident belongs to the system or to some specific person or set of conditions. Statistical methods provide the only method of analysis to serve as a guide to the understanding of accidents and to their reduction.

- Quoted from *Out of the Crisis,* on a program not soundly based:

  A program of improvement sets off with enthusiasm, exhortations, revival meetings, posters, pledges. Quality becomes a religion. Quality as measured by results of inspection at final audit shows at first dramatic improvement, better and better by the month. Everyone expects the path of improvement to continue to go along the dotted line.

  Instead, success grinds to a halt. . . . What has happened? The rapid and encouraging improvement seen at first came from removal of special causes, detected by horse sense. But as obvious sources of improvement dried up, the curve of improvement leveled off and became stable at an unacceptable level.

Then, [people] wake up; we've been rooked.

- Deming methods cannot be applied without an understanding of and the ability to apply statistical methods: Control Charts, Cause-and-Effect Charts, Run [Trend] Charts, Pareto Charts.

• • •

I believe that a safety professional whose practice was soundly based could be very comfortable as a member of a team applying Deming's principles. They emphasize the necessity of a culture change to achieve a significant improvement in quality; building quality into systems, in the beginning; a constant search for error reduction and reducing cycle time; a continuing improvement of processes and systems; placing responsibility for the greatest part of what can be done to attain improvement on management; not expecting from employees what they cannot do; and on a proper use of statistics.

Much of what Deming proposes represents a set of principles that I can easily support.

Philip B. Crosby sets forth markedly different concepts in *Quality is Free* (5). His program is prominently used and some safety professionals say that its adoption has done some good, but I wonder about its staying power. I also have some difficulties with it, as I do with hazards management programs that do not recognize the importance of design and engineering decisions, that do not emphasize the management aspects of continuing process and methods analysis and improvement, and that expect more of employees than they can accomplish.

While there are references to design in *Quality is Free,* the book's argument requires a stretch to conclude that product design and process improvement are all that important. Crosby presents a program outline very similar to many safety, health, and environmental programs, and a program of this sort probably expects more than can be achieved. It has several rah-rah aspects, with pledges and commitments from employees.

As the program proceeds, its focus becomes Zero Defects, a theme not now in vogue in very many places. The concept on which the Zero Defects program is built is of great concern to me. Crosby wrote:

> The Zero Defects concept is based on the fact that mistakes are caused by two things: lack of knowledge and lack of attitude.
>
> Lack of knowledge can be measured and attached by tried and true means. But lack of attention is a state of mind. It is an attitude problem that must be changed by the individual.

That premise seems focused on the employee's lack of knowledge and lack of attitude. Indeed, the program commences with Management Commitment, but senior management is not very prominent thereafter. It is also my impression that *Quality Is Free* expects too much of a quality assurance group and that the emphasis may be more on being in conformance than on reducing conformance ranges through improved product and process design.

Following is the Crosby program outline.

Quality Improvement Through Defect Prevention

STEP	PURPOSE
1. Management Commitment	To make it clear where management stands on quality
2. The Quality Improvement Team	To run the quality improvement program
3. Quality Measurement	To provide a display of current and potential nonconformance problems in a manner that permits objective evaluation and corrective action
4. The Cost of Quality	To define the ingredients of the cost of quality, and explain its use as a management tool

5. Quality Awareness	To provide a method of raising the personal concern felt by all personnel in the company toward the conformance of the product or service and the quality reputation of the company
6. Corrective Action	To provide a systematic method of resolving forever the problems that are identified through previous action steps
7. Zero Defects Planning	To examine the various activities that must be conducted in preparation for formally launching the Zero Defects Program
8. Supervisor Training	To determine the type of training that supervisors need to actively carry out their part of the quality improvement program
9. ZD Day	To create an event that will let all employees realize, through a personal experience, that there has been a change
10. Goal Setting	To turn pledges and commitments into action by encouraging improvement goals for themselves and their groups
11. Error-Cause Removal	To give the individual employee a method of communicating to management the situations that make it difficult for the employee to meet the pledge to improve

12. Recognition   To appreciate those who
                               participate

13. Quality Councils   To bring together the professional
                               quality people for planned
                               communication on a regular basis

14. Do It Over Again   To emphasize that the quality
                               improvement program never ends

Safety professionals would have an interest in certain aspects of a quality improvement program based on Crosby and would want to be active participants in them.

A somewhat different approach to quality management has been taken at Motorola through its Six Sigma Program, the goal of which is to achieve total customer satisfaction. The literature on Six Sigma states that defects in the manufacturing process are caused by insufficient manufacturing process control, by design margin, or by bad material. This places emphasis on product design and on the design and control of the manufacturing process.

What is Six Sigma? It represents a remarkable quality assurance standard. Envision a normal distribution chart, a bell curve. Variability from the mean, the center point of the distribution, is measured in units called Sigma, defined as the standard deviation. At plus or minus three Sigma, three standard deviations, 99.7 percent of a population would be included. In a manufacturing process, approximately 2,700 parts per million would fall outside of a plus or minus 3 Sigma variation.

At six Sigma, at plus or minus six standard deviations from the mean, it can be expected that no more than 3.4 parts per million will be defective. Four Sigma capability is ten times better than three Sigma capability; five Sigma is thirty times better than four Sigma; and six Sigma is seventy times better than five Sigma. One can appreciate the severity of the quality standard adopted.

What are the fundamentals of the Motorola program? In their literature, Motorola executives speak of:

- renewing emphasis on the basics
- participative management

- personal direction and involvement by senior management
- the tasks of creating a quality culture
- open and effective communication of ideas
- constant communication with the customer
- organization of teams that satisfy customer needs
- marketing, design, and manufacturing people forming teams to develop products that meet customer needs; products that can be produced quickly, reliably, and efficiently
- developing an integrated approach that enables designing in quality from the beginning
- using the entire range of statistical process controls to reduce variation

Paraphrased, Motorola "Six Steps to Six Sigma" are:

1. identifying the product or service you create or provide
2. identifying the customer and what the customer considers important
3. identifying your needs to create the product or service and to satisfy the customer
4. defining the process for doing the work
5. mistake-proofing the process
6. measuring and analyzing, thus ensuring continuous improvement in the process

Motorola has been exceptionally successful in its quality control endeavors. Its quality assurance courses are available to the public through Motorola University in Schaumburg, Illinois.

Federal Express was a 1990 winner of the Malcolm Baldrige National Quality Award. Fred Rine is much involved in the quality management program at Federal Express and is their Managing Director for Corporate Safety, Health and Fire Prevention. He wrote this about it:

> At Federal Express, we have developed our own Quality Development Plan. It includes portions of the criteria for The

Baldrige Award, some of Deming and some of Crosby, and elements comparable to those adopted at Motorola. The key to success is senior management support, commitment and involvement. Also, employees at all levels must be trained in "quality tools." Quality is here to stay. If companies are to be successful in the global market and service revolution, they must eliminate waste and do things right the first time. They must eliminate things that are not the "right things for the right reasons."

At Federal Express, each Director and above is required to develop a Plan for their specific functions utilizing the 7 Baldrige Quality Award Criteria/Items.

In a sense, this essay was intended to be a primer on certain quality assurance concepts and programs and to relate them to the basics of hazards management. I was led to this work by safety professionals engaged in quality assurance endeavors. They were right when they spoke of great similarities to hazards management concepts.

REFERENCES

1. 1992 Award Criteria, Malcolm Baldrige National Quality Award. United States Department of Commerce, Gaithersburg, Md.

2. W. Edwards Deming. *Out of the Crisis.* Cambridge, Mass.: Center for Advanced Engineering Study, Massachusetts Institute of Technology, 1986.

3. Rafael Aguayo. *Dr. Deming—The American Who Taught the Japanese About Quality.* New York: Carol Publishing Group, 1990.

4. Mary Walton. *The Deming Management Method.* New York: Putnam Publishing Company, 1986.

5. Philip B. Crosby. *Quality Is Free.* New York: McGraw-Hill, 1979.

Chapter 7

On Safety, Health, and Environmental Audits

I now believe that most safety, health, and environmental audits, intended to measure the quality of hazards management, are deficient in purpose and content. I include in that observation the many audits I have made and the many more audit reports I reviewed that were completed by my staff.

To those who make or will make audits or are responsible for their being made, I offer this self-review outline, which derives from an examination of the audit practices for which I was responsible, from a review of several other audit programs, and from a study of what has been evolving in the practice of safety.

1. What is the appropriate definition of an audit, and its purposes?

2. What is the *principal* purpose of an audit?

3. How is management commitment evaluated?

4. Should audits include an examination of the standards and practices for the design of the workplace, of work methods, and of environmental systems and controls?

5. Can assurance be given to management when an audit has been completed that the principal hazards have been identified and their attendant risks assessed?

6. How significant is it that recommendations be prioritized as to possible event consequence and probability?

7. Should management be given counsel as to alternatives with respect to eliminating or controlling identified hazards, the costs in relation to each alternative, and the amount of risk reduction expected?

8. Are audits directed to assist managements achieve their goals?

9. With what confidence can it be said that audits measure the actual quality of hazards management?

10. Should management expect that complying with recom-mendations contained in an audit report will result in a reduced frequency of incidents, and costs?

To begin with, I suggest that safety professionals agree on a definition of a safety, health, and environmental audit, and its purposes. Several writers have provided such definitions and stated what should result from an audit, and their views differ.

In *Safety Auditing* Donald W. Kase and Kay J. Wiese state: "Safety auditing, as we will develop it in this work, is a structured and detailed approach to reducing and controlling the seriousness of accidents. . . ." And, "Safety auditing is a method whereby management can receive a continuing evaluation of its safety effectiveness." (1)

In *Safety Audits: A Guide for the Chemical Industry* (2) issued by the Chemical Industries Association Limited, it is implied that the major objective of a safety audit is to determine the effectiveness of a company's safety and loss prevention measures. It is also proposed that whatever the objectives, it is important that they be clearly defined.

In an article titled "Measuring the Health of Your Safety Audit System," Robert M. Arnold, Jr., wrote: "Safety and health practitioners must properly diagnose the causes of safety and health problems before prescribing remedies." (3) Arnold lists seven benefits that a safety audit system should provide, including these two: "It offers a precise evaluation of an organization's safety performance," and "It establishes the organization's capability to forecast the potential for loss-producing events."

According to William E. Tarrants, editor of *The Measurement of Safety Performance* and author of several of its chapters: "One must recognize that the main function of a measure of safety performance is to describe the safety level within an organization . . . in effect, then, measures of safety performance must . . . describe where and when to expect trouble and must provide guidelines concerning what we should do about the problem." (4)

One could conclude from these definitions that a safety audit is a structured approach to provide a precise evaluation of safety effectiveness, a diagnosis of safety and health problems, a description of where and when to expect trouble, the capability to forecast the potential for loss-producing events, and guidelines concerning what should be done about the problems.

That's a laudable description. A safety audit that met all of those purposes would truly be professional and serve hazards management needs very well. But do our safety audits really do all that? I think that all of those purposes are seldom met.

What do I perceive to be the shortcomings? In the typical evaluation system, I don't believe enough attention is given to an organization's culture, to the reality of management commitment, to the significance of design and engineering practices and decisions, to the importance of seeking the possibility of the rare occurrences that result in severe consequences, to hazard identification and risk assessment, to prioritizing recommendations, and to giving counsel on alternative solutions and their costs.

An organization's culture determines the probability of success of a safety, health, and environmental program. All that takes place or doesn't take place in the program is a direct reflection of the organization's culture.

The paramount goal of a safety audit is to influence favorably the organization's culture. The safety auditor's principal intent should be to influence management to take a more proactive role in hazards management. A safety audit report, overall, would be an assessment of the culture. Even if management complied with every recommendation in an audit report, little would be gained if there was no change in its overall approach to hazards management.

Kase and Wiese drew the appropriate conclusion when they wrote: "Success of a safety auditing program can only be measured in terms of the change it effects on the overall culture of the operation, and enterprise that it audits." (1)

Management commitment is a part of and a reflection of the culture. And it may be that we don't properly measure management commitment. How is management commitment really manifested? Is the extent to which a results oriented accountability system impacts on the well-being of the management staff the principal measure of management commitment?

It is said that management measures what is important to it. That doesn't go far enough. There are at least two classes of measures—measures of form and measures of substance.

Measures of form are just that. Measures may be produced and observed, such as safety audits, meaningful incident statistics, or costs, and have no bearing on the status of those responsible.

If measures are of substance, there will be an accountability impacting on the well-being of those responsible for results, along with the other measures of performance. I would ask those who make safety audits whether, when commenting in reports on management commitment, they often give credit for something that isn't really there.

Evidence of an organization's culture and its management commitment is first demonstrated in its design and engineering practices, in its upstream decision making. Yet design and engineering practices are seldom addressed in typical audit systems, an exception being the audits that follow the guides prepared to meet the requirements of the Department of Energy and the Department of Defense.

System safety professionals have been trying for years to convince safety generalists that a superior level of safety performance cannot be achieved unless the upstream decision making properly addressed hazards and risks. And they got it right. In auditing a safety, health, and environmental program, the upstream addressing of hazards and risks should be considered next in importance, following culture and management commitment.

Joe Stephenson, in *System Safety 2000,* made strong comments on the importance of upstream design and engineering decision making. As I previously quoted, "The safety of an operation is determined long before the people, procedures, and plant and hardware come together at the work site to perform a given task."

He also stated:

> Safety professionals, managers, or supervisors who think they can have a significant impact on safety in the workplace by putting on a hard hat and safety shoes and meandering out with a clip-board to make the world safe are really fooling themselves if the upstream processes have not been properly done. Good safety practices must begin as far upstream as possible. (5)

If a safety audit does not result in influencing the upstream decision making as respects design and engineering practices and decisions, it can logically be expected that the same sort of adverse decisions will be made as are represented by the causal factors observed during the audit.

What would serve as a good reference for a safety professional who wanted to move toward including hazard identification and evaluation and risk assessment in the design and engineering decisions, both in the planning for and in the operation of a facility?

In "Safety in the Design Process"(Chapter 4), an outline was included of a policy statement for Process Design and Equipment Review. I suggest as an additional reference a section of *An Environment, Safety, and Health Assurance Program Standard* (6) which was prepared by Sandia Laboratories for the Department of Energy. That standard provides specifications for the establishment and execution of Environment, Safety, and Health (ES&H) Programs. There are five major sections in the Program: Management Support, Organization, Program, Line Organization ES&H Functions, and Staff ES&H Functions.

A synopsis follows of what appears under Line Organization ES&H Functions. Note that the functions are assigned to line management. Basically, they should be considered as elements of all safety, health, and environmental programs and the auditing practices related to them.

Planning
The line organization shall include ES&H considerations in the planning of all operations, and demonstrate through documentation that this has been done.

Facility/Project Status Information
Requires listing all facilities, projects, and activities that present significant hazards and classification by type of hazard.

Identification of Hazards and Their Risks
Significant hazards associated with each activity, project, or facility shall be identified and documented. The risk attendant upon each hazard shall be assessed and documented.

Requirements
ES&H requirements shall be correlated with identified hazards.

Risk Factors
Those factors that determine the risk shall be identified and controls established to adequately limit the risk. Control actions shall be documented and reported to management. The controls shall include those specified in the ES&H requirements and shall consist, as a minimum, of those applicable from the following:

Funding—Budget proposals shall assure that ES&H needs are taken into account.

Schedules—Adequate time is allowed for safe operation.

Facilities, Materials, and Equipment—The designs and specifications shall be reviewed to determine that ES&H considerations have been addressed, that hazards have been adequately controlled or eliminated, and that adequate facility environmental controls have been included.

Remaining categories in this section include the qualifications and training of people, written operating procedures, review of changes to determine ES&H effects, maintenance of records, corrective action, and compliance assurance, one feature of the latter being audits comparing performance to specifications.

These provisions presume that the culture and the management commitment include upstream design and engineering considerations and knowledge of hazard identification and evaluation and risk assessment methodologies.

Use of those methodologies will also be necessary in seeking out hazards that are seldom realized but which may result in severe injury or damage if they are realized. Such events rarely appear in incident records, yet they do occur. Hazards that are the causal factors—the possible unwanted releases of energy or of hazardous materials, presenting severe injury or damage potential—are not necessarily immediately obvious; indeed, they may be obscure. In the audit process, there should be a distinct attempt to seek out those obscure hazards.

An often-heard criticism of safety audits is that incidents resulting in significant injury or damage occurred after the audit was completed and that the contents of the audit report had no relation to the causal factors for those incidents.

Should management not logically expect that an audit report addresses the principal hazards and that their attendant risks have been assessed? Management's needs, those of employees, and those of the public are identical in respect to wanting to avoid incidents that result in significant harm or damage. To serve all three entities, an appreciation of Pareto's Law is necessary: about 20 percent of the incidents that occur will represent 80 percent of the problem.

Perhaps in making audits too much time is spent on the less significant and not enough time on identifying the major causal factors for low-frequency–high-consequence events.

If resources are unlimited and all of the recommendations submitted in an audit can be scheduled for corrective action immediately, there would be no need to prioritize them. That is never the case. A safety auditor has an obligation to give each hazard for which a recommendation is submitted in an audit report a severity of consequence and an occurrence probability indicator. Such indicators are included in "On Hazard Analysis and Risk Assessment" (Chapter 14) and are displayed in a Hazard Analysis/Risk Assessment Decision Matrix.

Would management be out of order if the safety auditor was asked these questions about the audit report? What are the most significant risks? On what do I concentrate first? Are there alternatives to what you have proposed? Can you help me with cost determinations for each of the alternatives and work with me to determine whether the risk reduction to be achieved for each of them is sufficient? I don't think so.

Too often safety auditors believe that they have done their jobs when an audit report is submitted containing a laundry list of recommendations, with no priority indications, with no offering of assistance. Audit systems fail if they do not show an understanding of management needs, if they are not looked upon as assisting in attaining management goals.

While audit reports must define hazards and actions needed to modify management systems, they will accomplish a great deal more if they are perceived as participative and supportive. How this audit function is accomplished has a great deal to do with effectiveness.

There is one category of audit, however, that deserves special comment because of the limitations of possibility and flexibility it permits. If the safety, health, and environmental audit is made exclusively to assist management in determining compliance with legally required standards, codes, or regulations, very little deviation from the "book" is permitted. It is my estimate that such an audit covers about 20 percent of program management needs.

OSHA has recognized that compliance with standards in itself is inadequate as a means to achieving effective safety and health program management. OSHA's Voluntary Protection Program states: "OSHA has long recognized that compliance with its standards cannot of itself accomplish all the goals established by the Act. The standards, no matter how carefully conceived and properly developed, will never cover all unsafe activities and conditions." (7)

I have wondered how safety professionals are perceived whose sole function is to obtain compliance with standards. That sort of thing has a very narrow range, unless, for occupational safety and health purposes, it is understood that the audit is also to speak to OSHA's General Duty clause, which requires employers to maintain a safe place to work, regardless of the existence of a standard.

But suppose the audit was made to fulfill the purposes of the composite definition of an audit that appeared earlier in this essay. Then the safety, health, and environment audit would provide a precise evaluation of safety effectiveness, a diagnosis of safety and health problems, a description of where and when to expect troubles, the capability to forecast the potential for undesirable events, and guidelines concerning what should be done about the problems.

If all that is done through the audit process, the auditor should be able to say to management with confidence that the actual quality of hazards management has been measured. After all, the audit will have given a precise evaluation, a diagnosis, direction as to where and when to expect trouble, and the capability to forecast undesirable events. If a safety, health, and environment audit is done properly, it seems that a safety professional should be able to state with some confidence that those purposes have been met. And that should be the goal.

I would like to consider the last item in the self-review list with which this essay commenced. Should management expect that complying with recommendations contained in the audit report will result in a reduced frequency of incidents, and costs? Sadly, the answer has to be—maybe. If the audit report is superficial, compliance with the recommendations it contains will, at best, achieve some short-term benefits. If the audit report accomplishes all of the purposes mentioned in the previous two paragraphs and is directed toward influencing the entity's culture, the answer would be yes—over time. An organization's culture is modified somewhat slowly.

Precise measures of safety, health, and environmental performance are difficult to obtain. Audits can be highly effective measures of the quality of hazards management if they are well conceived and well done.

REFERENCES

1. Donald W. Kase and Kay J. Wiese. *Safety Auditing: A Management Tool.* New York: Van Nostrand Reinhold, 1990.

2. *Safety Audits: A Guide for the Chemical Industry.* London: Chemical Industries Association Limited, 1977.

3. Robert M. Arnold, Jr. "Measuring the Health of Your Safety Audit System." *Professional Safety,* April 1992.

4. William E. Tarrants, ed. *The Measurement of Safety Performance.* New York: Garland Publishing, 1980.

5. Joe Stephenson. *System Safety 2000.* New York: Van Nostrand Reinhold, 1991.

6. *An Environment, Safety, Health Assurance Program Standard,* SAND79-1536. Prepared by Sandia Laboratories, Albuquerque, for the U.S. Department of Energy, 1979.

7. *OSHA's Voluntary Protection Program.* Federal Register, July 12, 1988.

Chapter 8

Defining the Practice of Safety

After participating with safety practitioners in what he considered a baffling discussion of concepts, a highly regarded professor observed that what we who call ourselves safety professionals actually do will never be accepted by those outside our field of endeavor as a profession until we agree on a clear definition of our practice. I agree with that premise, and will define the practice of safety in this essay.

It is a basic requirement of a profession to develop a precise and commonly accepted language that clearly presents an image of the profession. Terminology used by safety professionals should also convey an immediate perception of their practice.

In his book *General Insurance,* David L. Bickelhaupt made a significant statement about the need for clear communications that speaks to the purpose of this treatise: "Terminology becomes important in the serious study of any subject. It is the basis of communication and understanding. Terms that are loosely used in a general or colloquial sense can lead only to misunderstanding in a specialized study area such as insurance." (1)

Similarly, clear terminology and avoiding terms that lead to misunderstanding is necessary in the practice of safety, which surely is a highly specialized study area.

DEFINING SAFETY

First, safety must be defined as used in the phrase *the practice of safety*. If we cannot agree on a definition of safety, how could we assume that we are communicating with each other when we use the term? How could we assume that we communicate with those outside our profession if we use a multitude of definitions of safety? If those outside our profession don't understand our practice, it is unlikely that they would consider us professionals. And safety professionals should attain that recognition to encourage being sought for their counsel.

Dictionary definitions of safety are commonly given in safety literature. Since they are based on absolutes, such definitions are of little value to us. One dictionary defines *safety* as: "the quality of being safe; freedom from danger or injury," and defines *safe* as: "free from or not liable to danger; involving no danger, risk or error." Attaining the state—involving no danger, risk, or error—that would qualify for the definition of safety cited is not possible.

There is no such thing as an absolutely safe environment. That should be understood by safety professionals. If every known safety concept, standard, or regulation was applied in a given situation, there would still be the probability, however remote, of an injurious or damaging event occurring.

In the *1989 Annual Report* issued by the National Safety Council, safety was defined as "the control of hazards to attain an acceptable level of risk." (2)

In *Introduction to Safety Engineering,* Gloss and Wardle give this definition: "Safety is the measure of the relative freedom from risks or dangers. Safety is the degree of freedom from risks and hazards in any environment." (3) They state: "How safe is safe? Safety is relative—nothing is 100% safe under all conditions."

In *Occupational Safety Management and Engineering,* Willie Hammer wrote: "Safety—frequently defined as free from hazards. However, it is practically impossible to completely eliminate all hazards. Safety is therefore a matter of relative protection from exposure to hazards; the antonym to danger." (4)

Lowrance, in *Of Acceptable Risk: Science and the Determination of Safety,* stated: "We will define safety as a judgment of the acceptability of risk. . . . A thing is safe if its risks are judged to be acceptable." (5)

It is unrealistic to assume that an environment could exist in which the probability is zero of an injurious or damaging event occurring.

Borrowing from all whom have been quoted, this definition of safety is proposed as being applicable to the practice of safety and as being credible in dealing with decision makers:

Safety is a state for which the risks are judged to be acceptable.

Safety practitioners must attain a recognition that the term *safety* defines a relative and acceptable state for which the risks are judged to be tolerable. They must also be aware that implementing all that they propose, while serving to reduce risk significantly, will not eliminate risk entirely.

In the definitions of safety quoted, the terms *risk* and *hazards* are used, words frequently found in communications by safety professionals. In establishing what the practice of safety is all about, clear understanding is also necessary of these two terms.

DEFINING RISK

Arriving at a definition of risk applicable to the practice of safety that could be used convincingly in discussions with decision makers was not easy. Risk is a word that has too many meanings. Executives with whom safety professionals deal may hear the word used in several contexts in a given day.

Taking a business risk, a speculative risk, offers the possibility of gain or loss. That implies a meaning of risk different from that to which the practice of safety applies. Risks with which safety professionals are involved can only have adverse outcomes.

Definitions of risk in risk management and insurance literature emphasize uncertainty. They don't meet the purposes of the practice of safety. A few examples follow.

In *Principles of Insurance,* Mehr and Commach state: "risk is uncertainty concerning loss." (6) Greene defined risk in *Risk and Insurance* as "uncertainty that exists as to the occurrence of some event." (7) In Bickelhaupt's book *General Insurance,* he wrote:

"Basically, risk is uncertainty, or lack of predictability." (1) In *Risk Management and Insurance,* Williams and Heins define risk as "the variations in the outcomes that could occur over a specified period in a given situation." (8)

Risk fundamentally implies uncertainty. From a risk management and insurance viewpoint, it is understandable that risk is emphasized as being uncertainty. That's a basic actuarial concept. But the definitions quoted do not communicate entirely the nature of risk for which safety professionals give counsel. I cannot conceive of safety professionals presenting themselves solely as consultants in uncertainty reduction.

Also, the definitions of risk given in risk management and insurance literature seldom mention—even by implication—the severity of an event's consequences. Giving advice to reduce the possible severity of results of an incident is a significant part of the work of safety professionals.

Other authors include concepts of both probability and severity of consequences in their definitions of risk. In *Of Acceptable Risk: Science and the Determination of Safety*, Lowrance wrote: "Risk is a measure of the probability and severity of adverse effects." (5) Lowrance's definition of risk serves the purposes of the practice of safety exceptionally well. Rowe, in *An Anatomy of Risk,* gave this definition, which supports Lowrance's: "Risk is the potential for realization of unwanted, negative consequences of an event." (9)

And this definition appears in the section of the 1992 Federal Government budget titled "Reforming Regulation and Managing Risk Reduction Sensibly: "Risk is the individual likelihood (e.g., one accident in over nine million airline flights) of harm (e.g., airplane crash) from some level of exposure (e.g., 110 million airline flights 1975–85) to a hazard (e.g., windshear at airports). Risk and exposure are combined to determine how much harm (e.g., 12 crashes and 400 lives lost) is caused. The term 'risk' is also used to represent the sum of all individual risks, particularly when developing population estimates." (10)

Lowrance's definition of risk is recommended because of its simplicity and its completeness. It has been tested in real world safety practice, and it fits.

That definition requires both a measure of probability and a determination of the severity of adverse results. It promotes a thought process that asks: Can it happen? What is exposed to harm or damage? What will the consequences be if it does happen? How often can it happen?

Applying the definition of risk that appears in the Federal Government budget for 1992 (10) requires asking the same questions. Similar conclusions would result, which would be a measure of probability and severity of adverse effects.

In determining whether a thing, a situation, or an environment is safe, the decision making is judgmental. In the practice of safety, such a determination would follow an estimate of the probability of occurrence of an undesirable event and of its possible adverse consequences.

People make countless decisions to participate in activities for which they judge the risks to be tolerable. That judgement could be based on data derived from scientific study, or it could be almost entirely uninformed.

Studies made by scientists do not necessarily determine whether a thing is safe. Results of their studies will establish the probability of undesirable events under given circumstances and the severity of their outcomes. Determining whether a thing is safe requires a judgment of the acceptability of that probability and outcome. Scientists are no more qualified to make risk acceptance judgments than others. Judgments of risk acceptance may be made on a personal, economic, social, or political level. However those judgments are made, no situation is absolutely risk free.

DEFINING HAZARDS

Having defined risk, these questions should then be asked. What is the source of risk? What presents the probability of events occurring that could have adverse effects?

That source of risks is hazards. Hazards are defined as the potential for harm or damage to people, property, or the environment. Hazards include the characteristics of things and the actions or inactions of people.

Hazards are the justification for the existence of the entirety of the practice of safety, the purpose of which is the prevention or mitigation of harm or damage to people, property, or the environment.

Safety was defined as a state for which the risks are judged to be acceptable. Risk was defined as a measure of the probability and severity of adverse effects. All risks with which safety professionals are involved derive from hazards. There are no exceptions.

OUR BAFFLING AND NONDESCRIPTIVE TITLES

Whatever safety professionals call themselves, the generic base of their existence is hazards. If there are no hazards, there is no need for safety professionals. These statements apply, whatever words safety practitioners use in their titles—loss control, safety, risk control, environmental affairs, loss prevention, safety engineering, occupational health, and so on.

For many years, safety practitioners have been striving for professional recognition. Titles used by safety personnel may be a hindrance in their achieving an understanding of the practice of safety by those outside the profession. Safety practitioners call themselves by too many names, very few of which communicate an image of a recognized, professional practice.

An informal and unscientific study was conducted to assess how some of the titles safety practitioners use were perceived by management personnel. Risk managers were approached who had on their staffs people with titles like Director of Loss Control, Director of Loss Prevention, Industrial Hygienist, Safety Manager, Director of Safety, and Fire Protection Engineer. Risk Managers were asked to arrange a communication for me with their bosses or their bosses' bosses. Discussions were held to determine whether there was an understanding of the purposes of the people with the previously cited titles.

For the title Fire Protection Engineer, there was very good recognition of function and purpose. Unfortunately, the title Industrial Hygienist got the least recognition and was often equated with sanitation. As a part of a title, Occupational Health was frequently well understood as to role and purpose. Director of

Safety and Safety Manager as titles were quite well understood, though not as well as Fire Protection Engineer.

Loss Control and Loss Prevention as titles did not convey clear meanings and the recognition level as to function was poor. Loss Control was often believed to represent the security function of inventory control. On several occasions, Loss Prevention was assumed to be a part of claims management. Loss Control and Loss Prevention as functional designations have their origins in the insurance business. Within the insurance fraternity and among some other safety practitioners, the terms are understood. But those terms do not convey clear messages to those outside that group of users.

If I had a magic wand with which I could eliminate use of the titles Loss Control, Loss Prevention, and Industrial Hygienist by those engaged in the practice of safety, I would do so. And I would believe that I had performed a highly beneficial service. If the names safety professionals give themselves baffle people, how can what they do ever be considered a profession?

DEFINING THE PRACTICE OF SAFETY

Roger L. Brauer's definition of safety engineering in his book *Safety and Health for Engineers,* is a good place to start in developing a definition of the practice of safety: "Safety engineering is the application of engineering principles to the recognition and control of hazards." (11)

A compatible definition appears in Introduction to *Safety Engineering* by Gloss and Wardle: "Safety engineering is the discipline that attempts to reduce the risks by eliminating or controlling the hazards." (3)

At a meeting of the Board of Certified Safety Professionals (BCSP), a definition of safety practice was written during discussions of a project to validate that its examinations properly measure what safety professionals actually do: "Safety practice is the identification, evaluation, and control of hazards to prevent or mitigate harm or damage to people, property, or the environment. That practice is based on knowledge and skill as respects Applied

Engineering, Applied Sciences, Applied Management, and Legal/Regulatory and Professional Affairs."

Dr. Thomas A. Selders, CSP, CIH, PE, has said that the practice of safety should be anticipatory and proactive. Recognizing the definitions of safety practice and of safety engineering that I have cited, and Dr. Selder's comments, I propose the adoption of the following definition of the practice of safety.

The Practice of Safety

- serves the societal need to prevent or mitigate harm or damage to people, property, and the environment, which may derive from hazards
- is based on knowledge and skill as respects Applied Engineering, Applied Sciences, Applied Management, and Legal/Regulatory and Professional Affairs
- is accomplished through
  - the anticipation, identification, and evaluation of hazards, and
  - the giving of advice to avoid, eliminate, or control those hazards, to attain a state for which the risks are judged to be acceptable

The knowledge and skill required to enter the practice of safety and to fulfill the role of professional safety practice will be discussed in a later essay. Those knowledge and skill requirements are, of necessity, exceptionally broad. A list of the many functions performed by safety professionals would be quite lengthy.

Such a list could start with giving counsel on safety in the design process, and proceed into highly complex hazard analysis and formulating solutions for those hazards; having extensive knowledge of codes, standards, and regulations and serving as a resource concerning them; conducting training programs; producing incident analyses, making safety management audits—and one could go on for some time.

But consider this. No matter how extensive the knowledge and skill of the safety professional or whatever individual tasks are performed, is it not the end product of the safety professional's work to give advice to others?

Whatever the particular field of endeavor, the entirety of the practice of safety is represented in the definition I propose. It encompasses the work of Brauer (11), Gloss and Wardle (3), and the Board of Certified Safety Professionals. It properly relates to hazards as the generic base of the practice of safety. As defined, the practice of safety is to attain a state for which the risks are judged to be acceptable.

Why Safety Professionals Exist

Safety professionals exist for these purposes only:

- to anticipate, identify, and evaluate hazards,
- to give advice on the avoidance, elimination, or control of hazards, to attain a state for which the risks are judged to be acceptable

As defined, the practice of safety includes all fields of endeavor for which the generic base is hazards—occupational safety, occupational health, environmental affairs, product safety, all aspects of transportation safety, safety of the public, health physics, system safety, fire protection engineering, and so on. For all those for whom the generic base is hazards, a previously made statement applies: If there are no hazards, there is no need for their existence.

To fulfill their purposes, all hazards-related practitioners, regardless of the names they give themselves, would perform in accord with the definition of the practice of safety. And their practices are based on Applied Engineering, Applied Sciences, Applied Management, and Legal/Regulatory and Professional Affairs. (These four terms represent the major sections of the Comprehensive Practice Examination and the Specialty Examinations given by the Board of Certified Safety Professionals.)

Hazards Management—An Umbrella Designation

I recognize that the term "the practice of safety" is not commonly used and would probably not be easily accepted as encompassing all functions for which the generic base is hazards.

Thus, "hazards management" is offered as an appropriate umbrella designation for all practices for which the generic base is hazards.

It is the intent of this essay to define the practice of safety in a logical and precise language. All safety professionals who would like their pratice to be thought of as representing a profession are invited to move this discussion forward.

REFERENCES

1. David L. Bickelhaupt. *General Insurance.* Homewood, Ill.: Richard D. Irwin, 1983.

2. *1989 Annual Report.* Itasca, Ill.: National Safety Council.

3. David S. Gloss and Miriam Gayle Wardle. *Introduction to Safety Engineering.* New York: John Wiley & Sons, 1984.

4. Willie Hammer. *Occupational Safety Management and Engineering.* Englewood Cliffs, N.J.: Prentice-Hall, 1985.

5. William W. Lowrance. *Of Acceptable Risk: Science and the Determination of Safety.* Los Altos, Calif.: William Kaufman, 1976.

6. R. I. Mehr and E. Cammack. *Principles of Insurance.* Homewood, Ill.: Richard D. Irwin, 1976.

7. M. R. Greene. *Risk and Insurance.* Cincinnati: South Western Publishing Company, 1979.

8. C. Arthur Williams, Jr., and Richard M. Heins. *Risk Management and Insurance.* New York: McGraw-Hill, 1985.

9. William D. Rowe. *An Anatomy of Risk.* New York: John Wiley & Sons, 1977.

10. "Reforming Regulation and Managing Risk Sensibly," Section IX.C. Budget of the United States, Fiscal Year 1992.

11. Roger L. Brauer. *Safety and Health for Engineers.* New York: Van Nostrand Reinhold, 1990.

Chapter 9

Academic and Skill Requirements for the Practice of Safety

Having defined the practice of safety in the preceding essay, I would now like to discuss the knowledge and skill requirements for that practice.

As a beginning, seventeen text books commonly used in safety curricula were reviewed. Texts that focused on a particular and specialized subject fulfilled their purposes very well. Some of the texts emphasized safety management methods; others were based primarily on engineering practice. Four of the seventeen offered a good balance of safety engineering and safety management systems.

From even the best of the seventeen texts, it would be difficult to extract the two outlines I sought for this exercise, which are:

- *the academic knowledge and skills that would prepare one to enter the practice of safety, and*
- *the knowledge and skill requirements that describe the applied practice of safety, using the term broadly*

In fairness, the authors of the texts reviewed may not have intended to define the preferred academic accomplishments to enter the safety profession or to specify the knowledge and skill requirements of professional safety practice. Nevertheless, one can wonder about the value of certain commonly used texts as preparatory material for students taking safety courses.

Fortunately, the work of many others, far too numerous to mention, done through the American Society of Safety Engineers (ASSE) and the Board of Certified Safety Professionals (BCSP), serves as valuable resources. That work can be adapted in outlining both the academic knowledge that would prepare one to enter safety practice and the knowledge and skill requirements for professional safety practice.

JOINT ASSE AND BCSP BACCALAUREATE CURRICULUM STANDARDS

For quite some time, ASSE has undertaken to accredit college and university safety degree programs that met its standards. Many ASSE Board Members and others contributed to the development of a Suggested Core Curriculum for the Occupational Safety and Health Professional. That Suggested Core Curriculum has been used in the accreditation process.

In April 1980, BCSP issued a report authored by Ralph J. Vernon, Ph.D., CSP, titled *Curricula Development and Examination Study Guidelines* (1). A review had been made of the 4,400 applicants who had thus far been examined by BCSP, additional statistical studies were completed, and two research papers had become available from the National Institute for Occupational Safety and Health. One was titled *A Nationwide Survey of the Occupational Safety and Health Work Force* (2). The other title was *Development and Validation of Career Development Guideline by Task/Activity Analysis of Occupational Safety and Health Professions: Industrial Hygiene and Safety Profession* (3).

From those studies, Dr. Vernon prepared a Suggested Baccalaureate Curriculum for the Safety Professional (1). Also, a listing was developed of 348 Tasks/Activities and Performance Objectives that derived from the NIOSH studies. In August of 1981, BCSP published an extension of the original curriculum guideline in a Report prepared by The BCSP Ad Hoc Committee on Academic Guidelines and titled *Curricula Guidelines for Baccalaureate Degree Programs in Safety* (4). Both of these Reports remain exceptionally valuable as resources.

A review of the 348 Tasks/Activities and Performance Objectives listed by Dr. Vernon indicates that they are still applicable in defining a part of safety practice. As Dr. Vernon stated in a preface, the work and listings derived from the NIOSH studies "are generally limited to tasks performed by safety professionals in the context of preventing human injury and illness in an occupational setting."

Distribution by BCSP of those reports had a broad impact on many safety professionals, including me. At the time of their publication, I was a member of the Board of Directors of ASSE. Working with other Board members, modifications were proposed and made in the ASSE curriculum guidelines that moved them closer to those of BCSP.

Activity has now been completed to combine, in effect, the work of ASSE and BCSP into one curriculum guideline. ASSE has reached agreement with the Accreditation Board for Engineering and Technology (ABET) whereby ABET's Related Accreditation Commission (RAC) will assume responsibility for accrediting college and university safety degree programs, commencing in 1993. ASSE and BCSP have agreed on *Curriculum Standards for Baccalaureate Degrees in Safety* (5), which form the basis for the ABET/RAC accreditation criteria for safety degree programs.

Just how well would graduates be prepared to enter safety practice if they fulfilled the course requirements of those Curriculum Standards? In making that judgment, one would have to consider recent transitions in the responsibilities of safety professionals, the diversity of their roles, and the specialties within the practice of safety.

As an indication of transitions, the responsibilities of many safety professionals have been extended to include occupational safety, occupational health, and environmental affairs. Some also have involvement with fire protection, transportation safety, product safety, and other hazards related functions.

No college curriculum could include in a baccalaureate safety degree program all the courses that one would like students to take. So at a baccalaureate level the course work should be basic and preparatory and provide broad opportunities in anticipation of current and evolving professional needs.

All who have had a part in the development of what have become the *Curriculum Standards for Baccalaureate Degrees in Safety* should be commended for their work.

It is my opinion that the program it outlines serves well as a "standard" for the academic knowledge that would prepare one to enter the practice of safety. It is foundational and broad while appreciating the course requirements of individual universities and the necessity of having several electives in specialty fields.

I have made a thorough study of the suggested Curriculum Standards. In speculating on the ideal courses to be taken it was easy to get the course load up to 180 hours. And that's absurd. Each reviewer would no doubt suggest changes in requirements. And the discussions could go on forever.

And what do these Curriculum Standards propose? A brief review of their requirements follows.

Under "University Studies" minimum requirements, typically for lower level courses, are listed for subjects that usually exist in a variety of departments. Six subject areas represent knowledge and skill essential for safety practice.

- *Mathematics, Statistics, and Computer Science*
  Courses required: Calculus, Statistics, and Computer and Information Processing.

- *Physical, Chemical, and Life Sciences*
  Courses required: Physics (with laboratory), Chemistry (with laboratory), and Life Science (a laboratory component is recommended). An Organic Chemistry course is strongly recommended.

- *Behavioral Science, Social Science, and Humanities*
  Courses required: Psychology, and other Social Science and Humanities courses, totaling at least 15 semester hours.

- *Management and Organizational Science*
  Course strongly recommended: Introduction to Business or Management. Course recommended: Business Law or Engineering Law.

- *Communication and Language Arts*
  Courses required: Rhetoric and Composition, and Speech. A course in technical writing is strongly recommended.

- *Basic Technology and Industrial Processes*
  Courses required: Applied Mechanics, or its equivalent, and Industrial or Manufacturing Processes.

A program would be considered deficient if it did not contain a "strongly recommended" course, unless its absence could be well supported.

Next comes "Professional Core." Courses are to develop the basic knowledge and skills of the safety professional. Each of the following courses is required.

- Introduction to Safety and Health
- Safety and Health Program Management
- Design of Engineering Hazard Control
- Industrial Hygiene and Technology
- Fire Protection
- Ergonomics
- Environmental Safety and Health
- System Safety and Other Analytical Methods for Safety
- Experiential Education—there shall be an internship or COOP course

"Required Professional Subjects" are next listed, subjects that are important for safety professionals. However, it is not necessary that a full course be devoted to each subject, although a school has that option.

- Measurement of Safety Performance
- Accident/Incident Investigation and Analysis
- Behavioral Aspects of Safety
- Product Safety
- Construction Safety
- Educational and Training Methods for Safety

Then "Professional Electives" and "General Electives" follow in the Curriculum Standards. Thirty one possible elective subjects are listed as examples, with the indication that programs are not limited to the listing. General electives are to fulfill the additional course hours required.

It has been said that the ASSE and the BCSP curriculum guidelines represented an ideal program. That may have been so when they were first developed. Considering the recent transitions in the practice of safety and probable future requirements, it is difficult to find fault with the *Curriculum Standards for the Baccalaureate Degrees in Safety* as now written. They represent a sound course of study through which the needed academic knowledge can be acquired to prepare those who are to enter safety practice.

RELATIONSHIP BETWEEN THE ASSE/BCSP BACCALAUREATE CURRICULUM STANDARDS AND THE BCSP SAFETY FUNDAMENTALS EXAMINATION

The first level examination given by the BCSP is intended to test knowledge of Safety Fundamentals. Content of that examination should relate directly to the ASSE/BCSP Curriculum Standards if it represents a sound course of study. And it does.

Those board members who developed the *Safety Fundamentals Examination* (6) given by BCSP did great work. Concepts on which that examination are based were established long before I served on that board. Although the examination is regularly reviewed and updated in light of transitions in the practice of safety, its major themes remain the same. It is designed to measure the broad spectrum of *fundamental* knowledge required in the professional practice of safety.

"Basic and Applied Sciences" is the first of the six major sections in the *Safety Fundamentals Examination*. It represents approximately 33 percent of the examination. Each of the remaining five sections has about a 13 percent weighting. A brief review of its "Examination Content Definitions" indicates a close parallel with the ASSE/BCSP Curriculum Standards.

- *Basic and Applied Sciences*
  Mathematics, Physics, Chemistry, Biological Sciences, Behavioral Sciences, Ergonomics, Engineering and Technology, Epidemiology

- *Program Management and Evaluation*
  Organization, Planning, and Communication; Legal and Regulatory Considerations; Program Evaluation; Disaster and Contingency Planning; Professional Conduct and Ethics

- *Fire Prevention and Protection*
  Structural Design Standards, Detection and Control Systems and Procedures, Fire Prevention

- *Equipment and Facilities Design*
  Facilities and Equipment Design, Mechanical Hazards, Pressures, Electrical Hazards, Transportation, Materials Handling, Illumination

- *Environmental Aspects*
  Toxic Materials, Environmental Hazards, Noise, Radiation, Thermal Hazards, Control Methods

- *System Safety and Product Safety*
  Techniques of System Safety Analysis, Design Considerations, Product Liability, Reliability and Quality Control

KNOWLEDGE AND SKILL REQUIREMENTS FOR THE APPLIED PRACTICE OF SAFETY

Having discussed the academic knowledge that would prepare one to enter the practice of safety, consideration will now be given to the knowledge and skill requirements for the applied practice of safety, in a broad context.

An emphasis on "applied" is appropriate. In the definition given of the practice of safety, "applied" preceded engineering, sciences, and management. To define the "applied" practice of safety, the best reference found originated with the Board of Certified Safety Professionals.

BCSP gives five second-level examinations, which are called *Specialty Examinations* (6). They represent comprehensive practice and several specialties within the safety discipline. They also recognize that certain basic knowledge and skill are applicable to each discipline. All the specialty examinations contain the same major subject areas, but with different distributions, as shown in this chart.

BCSP SPECIALTY EXAMINATIONS
APPROXIMATE WEIGHTING OF SUBJECT AREAS

Examination	Engr.	Mgmt.	Applied Sciences	Legal/Regulatory & Prof. Affairs
Comprehensive Practice	35%	30%	20%	15%
Management Aspects	15%	45%	20%	20%
Engineering Aspects	45%	20%	20%	15%
System Safety Aspects	45%	25%	15%	15%
Product Safety Aspects	40%	25%	15%	20%

BCSP second-level examinations are "designed to test applied knowledge and the application of experience gained through professional practice and require a greater depth of knowledge than is required in the Safety Fundamentals Examination."

Fields of knowledge and skill represented in the examinations are exceptionally broad. It's doubtful that any safety professional would have deep knowledge and skill in all of the possible examination subjects. No one gets close to a perfect score. But each of the examination subjects represents a part of the practice of some safety professionals.

Examination domains and rubrics as now written have related well to the knowledge and skill requirements that describe the practice of safety. But some revisions can be expected as a result of a project in progress to validate that examinations properly relate to the recent transitions in safety practice.

There are four domains—Engineering, Management, Applied Sciences, and Legal/Regulatory and Professional Affairs. Brief comments follow on them and their knowledge and skill rubrics.

The *Engineering Section* is concerned with the application of the sciences for the safe and environmentally sound design of systems, processes, equipment, and products.

- *Safety engineering* is defined as engineering considerations related to analysis and control of injury-causing exposures and property damage. Knowledge fields include engineering and design methods, engineering mechanics, geotechnics/soil mechanics, structural systems, electrical systems, mechanical systems, materials handling, inspection and control procedures, ergonomics and human factors engineering, facilities planning and layout, and inferential statistics.

- *Fire protection engineering* is concerned with the application of fire protection engineering methods to the safeguarding of life and property against loss from fire, explosion, and related hazards. Knowledge fields include fire protection design parameters, fire detection systems, fire extinguishing systems, and process fire hazard control.
- *Occupational health engineering* addresses the application of engineering methods to analyze and eliminate or control exposures to environmental agents or stresses arising in work environments that may result in impaired health. Knowledge fields include process design parameters, industrial ventilation, noise control methods, radiation protection design parameters, and industrial waste control.
- *Product and system engineering* is concerned with the analysis, design, and manufacture of systems, processes, and products to identify, eliminate, or control hazards to the user of the product, system, or equipment. Knowledge fields include qualitative hazard analysis, quantitative hazard analysis, quality control and reliability, design parameters, and maintainability.
- *Environmental engineering* addresses the application of engineering methods to analyze and eliminate, control, or remedy hazards of environmental agents that may endanger the public, air, water, soils, or the environment in general. Knowledge fields include environmental engineering design, environmental analysis methods, waste management, control and remediation, air-quality management, and water-quality management.

The *Management Section* is concerned with the application of management principles and techniques to the management of safety programs and the safety function, including the relationships to health and environmental programs.

- *Applied management fundamentals* addresses the application of fundamental management principles to managing safety programs and the safety function. Knowledge fields include general principles of management, economic analysis, business law, and safety management techniques.

- *Business insurance and risk management* addresses the concepts and methods in business insurance and risk management that have direct application to safety. Knowledge fields include property and casualty insurance, workers compensation, product liability, and concepts of risk management.
- *Industrial and public relations* includes elements of industrial and public relations applicable to safety management. Knowledge fields include public relations, labor relations, personnel management, and safety training.
- *Organizational theory and organizational behavior* is concerned with the understanding of the design and function of various organizational structures and the behavior of individuals in organizations. Knowledge fields include organizational theory, types of organization structure, organizational behavior, and organization of the safety function.
- *Quantitative methods for safety management* addresses the application of quantitative approaches to the evaluation, analysis, and decision-making processes related to the management of the safety function. Knowledge fields include probability theory, descriptive and inferential statistics, computer science, and cost-benefit.

The *Applied Science Section* is concerned with the application of a knowledge of the *basic sciences and scientific principles* to the identification and evaluation of hazardous exposures. This section deals with *non-engineering applications of the basic sciences to safety problems.*

Knowledge fields include: chemistry, physics, mathematics, life sciences, and behavioral sciences.

The *Legal/Regulatory Aspects and Professional Conduct and Affairs Section* addresses safety legislation and associated standards and regulations, the process by which these are created and modified, liability considerations, professional ethics, and professional concerns.

- *Legal aspects* are concerned with understanding the impact of requirements imposed by safety, health, and environmental legislation, and the liability implications of tort and administrative law as applied to safety. Knowledge fields include legislative acts, liability, tort law and administrative law.

- *Regulatory aspects* are concerned with the promulgation and administration of regulation under legislative acts dealing with safety, health, and environment and with the concepts of consensus standards and their development and use. Knowledge fields include regulatory agencies, mandatory standards, and voluntary and consensus standards.

- *Professional conduct and affairs* addresses the ethical and legal concerns of the safety professional relative to society and colleagues. Knowledge fields include professional ethics, interpersonal relations, legal considerations, professional organizations, and significant developments, historical and current.

Without question, the contents of the BCSP specialty examinations represent a great breadth of "applied" knowledge and skill. And they realistically relate to the professional practice of safety.

Every recognized profession has developed a commonly accepted and precise language that clearly represents an image of a field of endeavor, and has established a body of knowledge that is unique to that profession.

It is the intent of this essay to

- present a unique course of study that would prepare one to enter the practice of safety, paraphrasing the *Curriculum Standards for Baccalaureate Degrees in Safety* approved by the American Society of Safety Engineers and the Board of Certified Safety Professionals

- give support to the validity of that course of study, by relating it to the Safety Fundamentals Examination given by the Board of Certified Safety Professionals

- present an outline of the knowledge and skill requirements that describe the "applied" practice of safety, adopting from the test specifications for the Specialty Examinations offered by the Board of Certified Safety Professionals

It is to the advantage of safety professionals, in seeking professional recognition, to promote an appropriate course of study for those entering safety practice and to establish broadly the standards for applied safety practice.

REFERENCES

1. Ralph J. Vernon. *Curriculum Development and Academic Study Guidelines.* Technical Report no. 1. Savoy, Ill.:, Board of Certified Safety Professionals, 1980.
2. *A Nationwide Survey of the Occupational Safety and Health Work Force.* National Institute of Occupational Safety and Health, Publication no. 78-164, July 1978.
3. *Development and Validation of Career Development Guidelines by Task/Activity Analysis of Occupational Safety and Health Professions: Industrial Hygiene and Safety Profession.* National Institute for Occupational Safety and Health, Contract CDC-99-74-94, 1977.
4. *Curriculum Guidelines for Baccalaureate Degree Programs in Safety.* Technical Report no. 2. Savoy, Ill.:, Board of Certified Safety Professionals, 1981.
5. *Curriculum Standards for Baccalaureate Degrees in Safety,* Joint Report No. 1. Des Plaines, Ill.: American Society of Safety Engineers, and Savoy, Ill.:, Board of Certified Safety Professionals, 1991.
6. Examination Information. Savoy, Ill.:, Board of Certified Safety Professionals, 1991.

Chapter 10

On Becoming a Profession

Safety practitioners continue to strive for recognition as a profession—within society, by other professions, by their employers, and from each other. They will attain that recognition only when safety practice meets the regimens of a profession, and only when the content and quality of their performance earn professional respect.

We who are engaged in the practice of safety use the word *professional* quite freely as a form of self-identification. For those who want to be considered safety professionals and who want the practice of safety recognized as a profession, a serious introspection concerning the status of what they do would serve their purposes well. An assessment is proposed to include:

- requirements for a profession
- advances made in the practice of safety in relation to those requirements
- practice content that earns respect and recognition as a profession
- additional inquiries and actions to advance the state of our art

This essay will present an assessment outline, which insightful readers could understandably debate.

Obviously, we must recognize with a sense of gratitude the accomplishments over several decades by those who have been successful in promoting a higher level of preparation for, and accomplishment in, the practice of safety. Many have contributed to that progress. A challenge remains for continued gains.

To begin with, it seemed appropriate to examine safety literature to determine what others have said about the status of what we do in relation to being a profession. Such writings are not extensive; most safety texts do not address the subject at all.

Two such articles on professional practice have appeared in *Professional Safety,* the magazine of the American Society of Safety Engineers (ASSE). In June 1981, Richard J. Finegan wrote "Is the loss control effort a profession?" (1). In November 1982, Dan Petersen's article, titled "Professionalism—a fourth step" began with this sentence: "Safety is working hard to become a profession." (2) Petersen suggested that we ought to examine our theoretical base by asking whether it's fact or opinion. Many of the questions he posed in 1982 remain valid.

In *Techniques of Safety Management,* Petersen set forth the need for introspection:

> In the safety profession, we started with certain principles that were well explained in Heinrich's early works. We have built a profession around them, and we have succeeded in progressing tremendously with them. And yet in recent years we find that we have come almost to a standstill. Some believe that this is because the principles on which our profession is built no longer offer us a solid foundation. Others believe that they remain solid but that some additions may be needed. Anyone in safety today at least ought to look at that foundation—and question it. Perhaps the principles discussed here can lead to further improvements in our approach and further reductions in our record. (3)

Chapter 1 of *Analyzing Safety Performance* (4), also written by Dan Petersen, is titled "The Professional Safety Task" and opens with a duplication of the *Scope and Functions of the Professional Safety Position* (5) published by ASSE. This publication is a job description, not a definition of a profession. It is divided into four major areas:

A. Identification and appraisal of hazardous conditions and practices and evaluation of the severity of the accident or loss problem.

B. Development of hazard control methods, procedures, and programs.

C. Communication of hazard control information to those directly involved, including the management, planning, and motivation necessary to integrate safety considerations into operations.

D. Measurement and evaluation of the effectiveness of the hazard control system and development of the modifications needed to achieve optimum results.

In 1980, Dan Petersen wrote this concerning ASSE's *Scope and Functions of the Professional Safety Position:* "While there is some discussion at the time of this writing as to whether or not this publication needs updating, there seems to be uniform agreement as to these four functions."

ASSE's *Scope and Functions of the Professional Safety Position* is a widely quoted document and has served its purposes well. Others have also suggested that it needs updating. Safety practitioners would benefit if that was done.

Ted S. Ferry made this brief mention of a profession in *Safety Program Administration for Engineers and Managers:*

> A profession is an occupation generally involving a relatively long and specialized preparation on the level of higher education and is usually governed by its own code of ethics. Nearly every profession has some safety and health aspects, some of them with distinct safety and health sub-disciplines. (6)

In *MORT Safety Assurance Systems* (7) by William G. Johnson, the chapter on "The Safety Function" early on quotes the ASSE *Scope and Functions of the Professional Safety Position.*

Introduction to Safety Engineering, by David S. Gloss and Miriam Gayle Wardle, contains the only reference found in a safety-related text that addresses the requirements of a profession:

Hallmarks of a Profession
If safety engineering is to be considered a profession, then it must meet the criteria for professionalization. Greenwood proposed that professions have specific characteristics.

1. A well-defined theoretical base

2. Recognition as a profession by the clientele

3. Community sanction for professionalization

4. A code of ethics, which regulates the professional's relationships with peers, clients, and the world at large

5. A professional organization

To define the current state of the profession, these limited references will be of some help, but not a great deal. To produce an outline for introspection, "Hallmarks of a Profession" cited from Gloss and Wardle will be extended considerably.

REQUIREMENTS FOR RECOGNITION OF THE PRACTICE OF SAFETY AS A PROFESSION

A. Establish a well defined theoretical and practical base, to include:

- a definition of the practice of safety
- the societal purpose of the practice of safety
- a recognized body of knowledge
- the methodology of the practice of safety

"Defining the Practice of Safety" (Chapter 8) was written as a beginning, to move the discussion forward, to set forth the societal purpose of the practice of safety, to speak of the rigor of education that would prepare one to enter safety practice, and to outline the methodology of safety practice.

As it was defined, the practice of safety includes all fields of endeavor for which the generic base is hazards—occupational safety, occupational health, environmental affairs, product safety, all aspects of transportation safety, safety of the public, health physics, system safety, fire protection engineering, et cetera. For all of these fields, if there were no hazards, there would be no reason for their existence.

The definition developed of the practice of safety is also given here as a reference for what follows.

The Practice of Safety

- serves the societal need to prevent or mitigate harm or damage to people, property, and the environment, which may derive from hazards
- is based on knowledge and skill as respects Applied Engineering, Applied Sciences, Applied Management, and Legal/Regulatory and Professional Affairs
- is accomplished through
  - the anticipation, identification, and evaluation of hazards
  - the giving of advice to avoid, eliminate, or control those hazards, to attain a state for which the risks are deemed to be acceptable

Having gone through the exercise of researching and writing "Defining the Practice of Safety," I concluded that the *Curriculum Standards For Baccalaureate Degrees in Safety* (9), jointly published by the Accreditation Council of the American Society of Safety Engineers and the Board of Certified Safety Professionals, represent a rigor of education to prepare one to enter the safety profession. Further, it was observed that the domains and rubrics of the specialty examinations (10) given by the Board of Certified Safety Professionals (BCSP) define the breadth of knowledge and skill required in applied safety practice.

Nevertheless, all of those observations are the first discussion subjects to be included in an introspection outline for moving state of our art toward recognition as a profession.

1. Shall we agree on, and promote an understanding of, a definition of the practice of safety and its basic methodology?
2. Do the *Curriculum Standards For Baccalaureate Degrees in Safety* jointly published by ASSE and BCSP represent sound preparation for one to enter the safety profession, and can we strongly promote their extended adoption?
3. Do the domains and rubrics of the BCSP specialty examinations represent the breadth of knowledge and skill required for safety practice and should we communicate that to the public and to safety practitioners?

Serious questions about our "recognized body of knowledge" continue to arise, since some safety practitioners still hold dearly to far too many myths. We should take a professional approach to examining them and, where appropriate, educate concerning them. Comments will be made here about only a few of those myths. Others can surely add to this list.

- At the ASSE Professional Development Conference in June of 1991, the keynote speaker informed attendees that 90 percent of accidents were caused by unsafe acts of employees. During sessions on Behavior Modification, similar statements were made. How pitifully unprofessional for safety practitioners to be so involved. Heinrich's 88-10-2 theory was held as the conventional wisdom years ago. It is a shallow myth.

- Many still offer as truth Heinrich's Foundation of a Major Injury—the 1-29-300 premise, which stated: "in a unit group of 330 similar accidents occurring to the same person, 300 will result in no injury, 29 will produce minor injuries, and 1 will cause a serious injury." Think about that—330 similar accidents occurring to the same person. Would that include falls off a fifty-story building?

  Bird and Loftus propose a different ratio: "1 disabling injury for every 100 minor injuries and 500 property damage incidents."

  Use of these statistical bases gives support to the principle that if we give adequate attention to the frequency of incidents, we will also be taking care of severity potential. That may or not be so, depending on whether the severity potential is also represented in the more frequent incidents.

  It has not been possible to locate a body of research that supports the validity of either the Heinrich or the Bird and Loftus postulations. They have the appearance of being mythical. Yet safety professionals continue to offer them as truths.

- Does the well-used axiom garbage in, garbage out—apply to our incident analysis systems? Do we mythically hold to information-gathering systems that produce misinformation?

- Is the premise that an unsafe act, an unsafe condition, and an accident are symptoms of *something wrong in the management system* simplistic and mythical? Does it serve professional needs? Can it be construed as being overly critical?

- In our concentration of efforts on occupational health, are we chasing a myth? Is the exposure really there? For well over forty years, we have been hearing about the latency period after which occupational illnesses would appear in workers compensation experience in great numbers. That has not happened. It's not suggested that there are not some real occupational illness case histories. But has the time come to wonder about the actual extent of occupational and environmental health exposures? Some scientists now question previously taken decisions and the validity of models used in arriving at health risk statements.

- Have we been reciting clichés, repeating the literature, without asking for substantiation? Do we docilely follow previously published premises, with no pretense at scientific inquiry as to foundation?

Furthermore, we may also have come up short with respect to "a recognized body of knowledge" for certain fundamentals:

- We use the terms *hazards* and *risks,* which are fundamental to defining our practice, in many contexts. Often, those with whom we try to communicate do not understand what we mean when we use those terms. And the literature on these subjects is baffling. There is a need for us to establish and agree upon meanings for *hazards* and *risks* and use them consistently in our communications.

- We could use some expositions on Pareto theory, adapted in many other fields as the 80-20 rule. Their purposes would be to discuss how an understanding of Pareto concepts might result in our becoming more professional time managers so that we would avoid, as far as is practicable, becoming bogged down in the less significant. Are we spending 80 percent of our time on 20 percent of the need? Is it not a necessity for a safety professional to be able to distinguish between the important and the less important? Basically, since all the resources, money, and time required to do it all will never exist, as professionals are we not required to be adept at priority setting?

- Little is written on the identification of severity potential as deserving its own place in a safety program. Those few who have proposed that the causes of low-frequency incidents resulting in severe consequences may be different from the causes of more frequent incidents deserve attention. This is an important subject for which we are not very proficient.

- We would benefit from some learned debate on the models for incident causation offered by a number of writers (McClay, Johnson, Benner, Petersen, De Joy, Haddon, Browning, et cetera). If we can not say that we understand incident causation, we are deficient in a fundamental need as professionals.

- Inquiries could profitably be made into the quality of incident investigations actually performed. They would delve into the requirements for and the probability of having effective systems, into the validity of the data obtained, and into the uses of that data.

- Information that would assist the safety professional as a participant in giving counsel to those who make design decisions on safety and health in the workplace or for products or equipment is almost nonexistent. Safety professionals are seldom participants in safety-related design concept discussions, and that's where hazards management starts. There is a great gap here as to safety practice, and immense opportunity.

- We could be much more effective in making safety audits, and need to expand our knowledge on how to be more professional in the process.

- Responsibilities of many safety professionals have been extended to include safety, health, and the environment. They would benefit professionally from some guidance as they enter their new fields.

- Many safety texts include information on hazard analysis and risk assessment. It's quite probable that safety professionals will have to become more adept in those elements of safety practice. A learned review that separates the often unattainable theory from realistic practice would be of value.

In no way was it intended that the preceding be all inclusive, especially coming from but one source. But, what has been written serves as the basis for the addition of the following discussion subjects to the inquiry list.

4. Should we examine safety literature to identify that which is patently unprofessional, with the intent of arranging exposition and debate on those subjects?

5. Should we be promoting adoption by safety professionals of methods of scientific inquiry and our seeking appropriate verification of that which is published?

6. Would we benefit from establishing a system to review the knowledge fields for which additional information is needed to maintain professional practice, and then arrange development and dissemination of that information?

B. Developing a common language within safety practice, with a realization that:

- a "hallmark" of a profession is to define itself
- the public does not know who we are or what we do
- we confuse others with our multitude of titles

Try this experiment: have your associates assume that they are asked by a member of the public what their job is—what do they do? I made such an inquiry with a group of safety professionals, and the responses were embarrassing. We have not established a common understanding of our practice, nor do we use a common language to define what we do. If we are to be recognized as a profession, we must be able to identify that which is unique about it, and its societal purposes.

One of the intentions in writing "Defining the Practice of Safety" (Chapter 8) was to emphasize the significance of our not having yet defined what we do. And to attain professional recognition, we must define our practice. Another purpose was to establish that the great variations in terms and titles we use may confuse those with whom we try to communicate.

Names of other professions—law, medicine, for example—immediately bring to mind a mental image of that field of endeavor, and the requirements to be a member of the profession. And we must follow that lead and define who we are. What we call ourselves and the language we use in communications both in the community at large and with our clients should convey the same sort of image of a discipline. We have a long way to go. There is no question that we baffle decision makers, and the public, with the multitude of titles we give ourselves.

A brief and unscientific study was made of decision makers' understandings of the titles we use. I then wrote that if I had a magic wand with which I could eliminate the use by safety practitioners of the titles Loss Prevention, Loss Control, and Industrial Hygiene, I would do so and believe that I had done a great service for those engaged in the practice of safety.

In addition, I said that the generic base of the practice of safety is hazards, and that if there were no hazards, there would be no reason for safety professionals to exist. In time, we will more than likely be considering an umbrella term that encompasses all aspects of safety practice. Hazards Management would serve very well for that purpose.

These are, then, additional discussion subjects.

7. Shall we undertake an exercise toward developing a commonly accepted language that clearly presents an image of the field of endeavor that is the practice of safety?

8. Do we arrange for a study of the public understandings of the titles we use, then promoting the use of those which best convey the image of a profession.

C. Achieving recognition as a profession by the clientele to whom we give advice, considering:

- the content, the substance, of the advice we give
- the nature of our communication

A part, but not all, of what we would do to achieve recognition by our clients as professionals has been previously addressed. It's possible that the substance of the communications of too many of us to decision makers is perceived as shallow and superficial and not pertinent to an entity's real hazards management needs. Much of our language developed years ago. It is time to evaluate the real substance of most of it. As an example—have the concepts of unsafe acts and unsafe conditions run their course?

Of greater concern is the improbability of clients considering what we do as professional if we insist on a separateness, of having purposes that are not perceived as helping in their attaining their goals, and on using language that may be considered accusatory.

The safety professional should strive to be perceived as part of the management team and as an individual who understands the goals of and the constraints on the organization.

Yet our literature frequently indicates that all incident causation derives from something wrong in the management system. And we promote the idea, over and over. People don't like to be constantly reminded of their shortcomings. Is the premise too simplistic? Through its use, are we being overly critical of a group of which we want to be a part? Do we gain more or lose more in its never-ending use? Yes, it can be theorized that there was a management shortcoming for every hazard, the realization of which resulted in, or may have resulted in, injury or damage. And it's necessary to identify those shortcomings in causal studies.

But it's probable that more frequent acceptance by our clientele can be obtained through language that demonstrates participation toward achieving goals we share, and that we can do so without language that seems to be overly critical.

At our professional conferences, we often discuss "how to obtain management support." Those discussions usually encompass methods of communication that might be more effective, and that's important. But I also suggest that we might gain from an examination of the content of our practice and of the substance about which we communicate. If our practice is not considered to be professional, the world's best communicator will still not get the management support and involvement necessary.

What is being proposed is that we explore these action subjects.

9. Should we examine the validity of incident causation theory, with the purpose of assuring a professional content in giving advice to our clientele?

10. Do we benefit by exploring how the terminology in our literature and that used by safety professionals is perceived by our clientele?

D. Promoting and supporting research, recognizing that:

• knowledge requirements concerning hazards and the methodologies to anticipate, eliminate, or control those hazards will continue to expand

• it is typical of a profession to seek to innovate, and continuously test previously established premises

How much needs to be said about the additional knowledge needs of safety professionals, especially in the past ten years? Is it true that the half-life of a new engineer is only seven years? It should concern us that safety research is most often done by people who would not necessarily consider themselves to be safety professionals. We are almost entirely excluded from determining what the research needs are or from assessing the results.

A proposal has already been made about the desirability of instituting a program to evaluate the premises that have accumulated in the practice of safety in the past seventy-five years. This is a subject for which activity by safety practitioners, with rare exceptions, will be an original undertaking.

This discussion subject is proposed:

11. Should safety professionals take the initiative in arranging, promoting, and supporting research projects on an expanding basis?

E. Maintaining rigid certification requirements, promoting the significance of certification, and giving additional status to certification

Much thanks should be given to those visionaries on the Board of Directors of the American Society of Safety Engineers in the 1960s who conceived of and established the Board of Certified Safety Professionals. As a result of their work, a sound and proven certification program exists for safety professionals, for which recognition continues to grow.

Very little will be said here about the BCSP program, which has stood the test of time. Numbers of those seeking examinations and certification have been increasing. In its determination to be current, BCSP has undertaken a validation project to assure that its examinations properly examine the practice of safety as it has evolved. Almost certainly, examination rubrics will be expanded.

All involved—individuals, the BCSP, and the founding group, ASSE—could do a better job of promoting the significance of certification and of giving a higher status to it. Individuals who employ safety professionals especially can give greater significance to the CSP designation. ASSE has strongly publicized certification and BCSP examinations and has organized examination preparation courses.

A discussion subject follows:

12. Should individual safety professionals, BCSP and ASSE be further promoting the significance of obtaining professional designation as a Certified Safety Professional?

F. Adhering to an accepted standard of conduct, which is an absolute requirement of a socially recognized profession

How can anyone presume to be a professional without being willing to meet high standards of performance and to insist that others in the profession do the same? A prescribed statement of professional conduct should exist covering the relationships expected in individual practice, in relations with one's peers, in dealing with clients, and with the community at large.

This discussion subject is appropriate:

13. Should safety professionals promote the publication of an agreed-upon and current statement on a professional standard of conduct, publicize it, and encourage safety professionals to consider it as foundational in their own practices and in their dealings with others?

G. Having a professional society, participating in it, and supporting it

Several related societies exist that safety professionals can well support. Low levels of participation in such societies lead to the observation that more safety practitioners call themselves safety professionals than should. Just being a member doesn't qualify for professional status insofar as performance is concerned.

To those who seek professional status and would like their practice to be recognized as a profession, this question is appropriate: should they be more prominent participants in the societies of which they are members?

For discussion, this is offered:

14. How do those safety professionals who want recognition as professionals become convinced that they should be active participants in their professional societies?

H. Obtaining societal sanction for professionalization

This is a futuristic goal. It will have been achieved when the public perceives that what those who designate themselves as safety professionals do has a distinct value to society. Safety professionals will earn that respect and recognition only through their performance.

When that occurs, it will be expected that a person with a prescribed professional education, experience, and certification will fulfill safety responsibilities. It is assumed that a doctor is a licensed physician, and that a lawyer has passed the bar examination. We've come a long way and yet have a long way to go, especially with the general public. At least in employment, there has been a growing recognition of the value of certification, a form of societal sanction. There has been a continuing increase in the percentage of advertisements for safety positions that mention the desirability of being a Certified Safety Professional.

Accomplishment for each of the action subjects in this outline will serve to achieve community sanction of professional status for those engaged in safety practice, which is an ultimate ideal.

REFERENCES

1. Richard J. Finegan. "Is the Loss Control Effort a Profession?" *Professional Safety,* June 1981.
2. Dan Petersen. "Professionalism—a Fourth Step." *Professional Safety,* November 1982.
3. Dan Petersen. *Techniques of Safety Management.* Goshen, N.Y.: Aloray, 1989.
4. Dan Petersen. *Analyzing Safety Performance.* New York: Garland STPM Press, 1980.
5. *Scope and Functions of the Professional Safety Position.* Des Plaines, Ill.: American Society of Safety Engineers.
6. Ted S. Ferry. *Safety Program Administration for Engineers and Managers.* Springfield, Ill.: Charles C. Thomas, 1984.
7. William G. Johnson. *MORT Safety Assurance Systems.* New York: Marcel Dekker, 1980.
8. David S. Gloss and Miriam Gayle Wardle. *Introduction to Safety Engineering.* New York: John Wiley & Sons, 1984.
9. Joint Report No.1, *Curriculum Standards For Baccalaureate Degrees in Safety.* Joint Publication of the Accreditation Council of the American Society of Safety Engineers (Des Plaines, Ill.) and the Board of Certified Safety Professionals (Savoy, Ill.), August, 1991.
10. Examination Information. Board of Certified Safety Professionals. Savoy, Ill., 1991

Chapter 11

On Causation Models for Hazards Related Incidents (HAZRINS)

Safety practice is accomplished through the anticipation, identification, and evaluation of hazards, and the giving of advice to avoid, eliminate, or control those hazards to prevent or mitigate harm or damage to people, property, or the environment. Advice given, if applied by the decision makers, should result in attaining a state for which the risks are judged to be acceptable.

Professional safety practice, therefore, requires fundamental knowledge of hazards and of how, through their realization, they become causal factors in hazards related occurrences.

Safety professionals investigating a hazards related incident should identify the same causal factors, with minimum variation. That is unlikely if their thought processes have greatly differing foundations.

As Ludwig Benner, Jr., wrote in "Rating Accident Models and Investigation Methodologies," a study that was undertaken for OSHA: "The number of conceptual accident models that drive government accident investigation programs seems unnecessarily diverse. Since they conflict, all models can not be valid." (1)

Advice given by many safety practitioners in or not in government is based on several of those diverse and conflicting models, all of which can not be valid.

In moving the state of their art forward, safety professionals would benefit greatly from exposition and debate on previously published models pertaining to the accident phenomenon, on their validity, and on what the bases should be of the models they use. Previously published models represent a great divergence of thinking:

This quotation from a letter received from Benner gives one indication of that divergence of thinking.

> Accident models and accident causation models involve two different areas of endeavor. The point is subtle, but in my view it is absolutely imperative to recognize the difference. Causation models purport to present cause and effects without identifying the phenomenon; no beginning and end of the phenomenon is indicated. Accident models, on the other hand, deal descriptively with accidents as a process that has a beginning and an end, and the elements of that process. Please help me keep my models in the latter arena when quoting any of my work to ensure that it is not thrown into the causation model arena inadvertently.

Respectfully, an attempt will be made to comply with Benner's request. But several references will be made to Benner's work in this essay since it is considered to be important, in contemplating accident models or accident causation models.

As used in this treatise, causation means the act or agency of causing or producing an effect. Also, it will be presumed that an appropriate causation model for hazards related incidents (HAZRINS) will treat an incident as a process and require a determination of when the phenomenon begins and ends.

A literature review indicates that several authors have recognized, with some frustration, the absence of a generally accepted accident causation model, and a variety of causation concepts and models have been proposed. Robert E. McClay established the need for an accepted causation model in "Toward a More Universal Model of Loss Incident Causation":

> The most obvious example of a weakness in the theoretical underpinning of Safety Science is the lack of a satisfactory explanation for accident causation. . . . Line managers in an organization can be forgiven for being cynical about safety when the reasons for the

occurrence of accidents seem so obscure. . . . What is needed is an acceptable model that explains the occurrence of accidental losses of all types across the entire discipline of Safety Science. (2)

In *MORT Safety Assurance Systems,* William G. Johnson states: "Improved models of the accident sequence would be helpful in understanding the dynamics of accidents and would be a basis for data collection. No fully satisfactory model has yet been developed, but many are promising and useful." (3) Ted S. Ferry, in *Modern Accident Investigation and Analysis,* wrote: "The scientific literature on mishap analysis offers little insight into the process by which mishaps occur." (4)

Variations in current practice and the need for an accepted investigation methodology are mentioned by Benner in "Accident Investigations: Multilinear Events Sequencing Methods":

> Approaches to accident investigations seem as diverse as the investigators. . . . The absence of a common approach and differences in the investigative and analytical methods used have resulted in serious difficulties in the safety field . . . including . . . barriers to a common understanding of the phenomenon . . . (and) popular misconceptions about the nature of the accident phenomenon. . . . The purpose of this paper is to call attention to the need to develop generally acceptable approaches and analysis methods that will result in complete, reproducible, conceptually consistent, and easily communicated explanations of accidents. (5)

One of the difficulties to be overcome in establishing a causation model is the determination of what it is to encompass and what terms are to be used in the description. As McClay wrote:

> . . . An ideal model should be applicable across the full spectrum of Safety Science. . . . It becomes necessary to use a broader term than "accident causation". . . . The term loss incident will be used . . . to include any event resulting from uncontrolled hazards, capable of producing adverse, immediate or long term effects in the form of injury, illness, disability, death, property damage or the like. (2)

Safety professionals give many names to the incidents to which a causation model would apply—accidents, incidents, loss incidents, mishaps, near-misses, occurrences, events, illnesses, fires, explosions, windstorms, drownings, electrocutions, et cetera. Pat Clemens, a prominent safety consultant, has said that the language used by safety practitioners lacks words to convey precise and understood meanings. Perhaps we should create our own words for that purpose. As an example, HAZOP (for Hazard and Operability Study) has become a commonly used and understood word rather quickly within the safety profession.

As stressed before, if there were no hazards, there would be no need for safety professionals. Hazards are defined as the potential for harm or damage to people, property, or the environment. Hazards include the characteristics of things and the actions or inactions of people. If a hazard is not avoided, eliminated, or controlled, the *potential will be realized.* Whatever names we use to identify those realizations—all of the types of incidents mentioned in the previous paragraph—they are all hazards related. There are no exceptions.

So it is proposed that a new name be created to encompass all hazards related incidents—HAZRINS—and that there be extensive discussion about an appropriate framework for thinking about a HAZRIN Causation Model.

Rather than emphasize the outcome, emphasis should be given to the processes through which significant adverse results may be attained. Safety literature commonly gives different names to incidents depending on their outcomes, such as "incidents" or "near-misses" for those that do not result in injury, and "accidents" for those that do. What is important is the *possible consequence.*

HAZRIN as a term encompasses all incidents that are the realization of the potential for harm or damage, whether harm or damage resulted or could have resulted—for all fields of endeavor that are hazards related. The people with whom safety professionals try to communicate are probably baffled by the many terms used to describe hazards related incidents.

An essay of this sort cannot encompass all of the accident models and accident causation models previously published, of which there are at least twenty. Two themes will be addressed in the remainder of this treatise.

First, a review will be given of the history and content of the Domino Sequence causation model established by H. W. Heinrich, which focuses on "man failure." This model, and the several causation models that are extensions of Heinrich's theorems, have dominated the thinking of a great many safety practitioners. I will establish that advice based principally on Heinrich's theorems is inadequate.

Then, a HAZRIN causation thought process will be outlined, reflecting my impressions of what is evolving in the practice of safety. It will:

- consider Haddon's energy release theory as fundamental in dealing with causal factors

- extend Haddon's energy release concepts to include the release of hazardous materials

- treat a HAZRIN as a complex process with an identifiable beginning and end

- recognize that multiple events, occurring sequentially or in parallel, in time and influencing each other, may precede the incident that results in, or could have resulted in, injury or damage

- establish the significance of an organization's culture in relation to hazards related incidents

- propose that management commitment to hazards management, deriving from its culture, should stand separately, and that it governs all other aspects of hazards management

- give design and engineering considerations distinct and primary status, with an emphasis on ergonomics

The third edition of *Industrial Accident Prevention* (6) by H. W. Heinrich, published in 1950, is the source of the following:

> In the middle 1920's, a series of theorems were developed which are defined and explained in the following chapter and illustrated by the "domino sequence." These theorems show that:
>
> (1) industrial injuries result only from accidents,
>
> (2) accidents are invariably caused by the unsafe acts of persons or by exposure to unsafe mechanical conditions,

(3)   unsafe actions and conditions are caused by faults of persons, and

(4)   faults of persons are created by environment or acquired by inheritance.

From this sequence of steps in the occurrence of accidental injury it is apparent that man failure is the heart of the problem. Equally apparent is the conclusion that methods of control must be directed toward man failure.

Figure 2 on page 11 of the text displays the "domino sequence." Later, this description of the "domino sequence" appears:

The several factors in the accident occurrence series are given in chronological order in the following listing:

1. Ancestry and Environment

2. Fault of person

3. Unsafe act and/or mechanical or physical condition

4. Accident

5. Injury

One of the principle Heinrich premises (taken from page 18) is often still cited by safety practitioners: ". . . a total of 88 per cent of all industrial accidents . . . are caused primarily by the unsafe acts of persons."

For years, many safety practitioners based their work on Heinrich's theorems, working very hard to overcome "man failure," believing with great certainty that 88 percent of accidents were caused primarily by unsafe acts of employees. How sad that we were so much in error. (I still have my set of "dominos," which are over forty years old.)

Heinrich's premises, and the several causation models that are based on them, are still the foundation of the work of many safety practitioners.

Indeed, most causation models have focused on the behavior of the individual presumed to have acted unsafely, rather than on the design of the work environment and on the design of work methods. And many safety practitioners, in the prevention measures they have proposed, have emphasized training programs, quality of leadership by supervisory personnel, behavior modification, appropriate methods of discipline—a great range of activities directed toward the control of "man failure." They are intended to achieve a change in the performance of this poor employee who, when judged retrospectively, is deemed to have acted unsafely.

Several modifications of Heinrich's "domino sequence" have been made, although they continue to focus on unsafe acts as the principal accident causes and thus on the behavior of the individual affected. Some of those models now include a reference to the "management system" and suggest that the so-called unsafe behavior may have been "programmed."

An example of that sort of thinking is contained in *Techniques of Safety Management* (7) by Dan Petersen, who lists "The Ten Basic Principles of Safety," beginning with: "An unsafe act, or an unsafe condition, and an accident are all symptoms of something wrong in the management system." Petersen's seventh principle is "In most cases, unsafe behavior is normal behavior; it is the result of normal people reacting to their environment. Management's job is to change the environment that leads to the unsafe behavior."

Petersen also stated that "Unsafe behavior is the result of the environment that has been constructed by management. In that environment, it is completely logical and normal to act unsafely."

In the context given, just what is unsafe behavior, if it is logical and normal? Very few of the authors who use the term *unsafe act* define it, and no attempt will be made to do so here.

An incident analysis, based on a causal model that has unsafe acts of the individual performer as the focus will often result in the wrong advice being given to decision makers. Such analyses miss the fundamental question—what aspects of workplace and work methods design are involved. They most often result principally in recommending some sort of behavior modification to correct "man failure."

If the environment that has been constructed—which would include both the design of the physical aspects of the workplace as well as the design of work methods—requires logical and normal behavior that is considered unsafe, it would seem that a HAZRIN model should begin with and focus on environmental causal factors. Doing so would place the emphasis of safety practice on origins, rather than on results such as unsafe acts considered to be "man failure."

Safety practice, then, commences with design and engineering considerations. P. L. Clemens and R. R. Mohr, in a course called *Risk Management and Aerospace Research* (8), support that premise. They wrote that the *order of effectiveness* of hazards management measures is: design, engineered safety features, safety devices, warning devices, and procedures/training.

Phrases such as "something wrong in the management system," commonly found in causation models require comment. What does the phrase mean? Does it belong to safety practitioners exclusively? Does its use do us more harm than good? From what particular posture is the term used? Is it offered from a distance or as a part of the management team? Is it accusatory?

Within any enterprise, each instance of less than desired performance, regardless of its nature, can be an indication of "something wrong in the management system." Safety professionals do not "own" the phrase. Use of such a causative indicator does not add to the uniqueness of safety practice. Surely, when addressing causative factors after the fact, it is appropriate to consider the management decisions previously made. But do safety professionals need to wear "something wrong in the management system" on their sleeves, as if to say "you management people blew it", over and over again. Are safety professionals not one of the "management people"?

A hazards related incident causation model should emphasize the special nature and the uniqueness of causation. It should promote an anticipatory thought process and be directed toward giving advice intended to avoid work environments "in which it is completely logical and normal to act unsafely."

Dr. Alphonse Chapanis (9) refers to situations of this sort as "error provocative." If what has been designed as a workplace or work method is "error provocative," how could anyone logically conclude that the principal incident cause was "man failure," meaning the unsafe act of the employee involved?

I have written that the greatest strides forward as respects safety, health, and the environment will be made in the design and engineering processes. That includes the design and engineering of facilities, processes, work stations, and tools and, very importantly, the design of work methods, the design of the procedures for accomplishing the work. And I continue to believe that to be so. An appropriate HAZRIN model would give emphasis to the anticipation of hazards early in the design and engineering process.

As the concepts of ergonomics/human factors engineering are more prominently applied, they will have an impact on professional safety practice in several ways. One such impact will be the shifting of emphasis from the actions of the individual, the so-called unsafe act, to considerations of workplace and work methods design.

As an example, it has become apparent that for many ergonomics related injuries and illnesses, for which the causes were classified as unsafe acts of employees in accident analyses, and for which behavior modification measures were proposed as corrective actions, the root causes were fundamentally matters of workplace and work methods design.

It's more than probable that the conclusions pertaining to unsafe acts drawn from injury and illness causal analyses fits very well under the much used axiom "garbage in, garbage out."

A causation model that focuses on the characteristics of the individual, on unsafe acts being the primary causes of workplace accidents, on corrective measures being directed principally toward effecting the behavior of the individual, on correcting "man failure," and on "predispositions, motivations, and attitudes" of the individual, results in a shallowness of safety practice, much of which may be misdirected and ineffective.

David M. DeJoy, in "Toward a Comprehensive Human Factors Model of Workplace Accident Causation," alluded to the inadequacy of efforts that stopped with the "predispositions, motivations, and attitudes of workers":

> The essence of the contribution of human factors to safety is that machines, equipment, jobs, processes and environments can be

safer if they are designed with the capabilities and limitations of the worker in mind. While the predispositions, motivations, and attitudes of workers are important to safety performance, a comprehensive human factors analysis goes well beyond these considerations. (12)

Key phrases in the foregoing paragraph, in the context of this paper, are "designed with the capabilities and limitations of the worker in mind" and "a comprehensive human factors analysis." They suggest an anticipatory approach, in the design phase, and applying human factors concepts, going beyond "predispositions, motivations, and attitudes," which are basic in the Heinrich theorems and in the causal models that are extensions of them.

And focusing on the errors of the individual won't necessarily lead to problem identification. Mark Paradies gives this view in "Root Cause Analysis and Human Factors":

> . . . Using the word error (as in operator error) tends to focus attention on the individual involved rather on the problem. This was driven home one day when an investigator said, "I can't list this as operator error—the operator wasn't at fault, the procedure was written wrong! "(10)

Note the phrase "the procedure was written wrong!" A focus on errors in a *process* can be fruitful. That is a major element in Total Quality Management and in "concurrent engineering." Focusing on improving the process is much more effective than concentrating on individual behavior.

Is the 85-15 Rule (11) attributed to W. Edwards Deming, world-renowned for his capabilities in quality assurance, also applicable in safety? He states that 85 percent of the problems in any operation are within the system and are the responsibility of management, while only 15 percent lie within the worker.

Professional safety practice requires that the advice given be based on a sound hazards related incident causation thought process so that in the application of that advice an effective utilization of the available resources can be attained.

It's proposed that safety professionals examine the premises, the causal models, on which their practices are based. In that exercise, consideration should be given to the desirability of removing "unsafe acts and unsafe conditions" from their language as causal factors. *All events deriving from things or from the actions or inactions of people that contribute to a hazard related incident are causal factors.*

Safety professionals might also want to get rid of their "dominos," which are overly simplistic representations of incident causation. (At a recent meeting at the National Safety Council, it was said that for more than ten years it has been the intent not to use the term "unsafe act" in their literature.)

For those who would undertake such an exercise in the development of sound premises concerning hazards related incidents (HAZRINS), the following are recommended as minimal readings:

- *Investigating Accidents with STEP* (13) by Kingsley Hendrick and Ludwig Benner, Jr.
- Other papers by Benner (1, 5, 14)
- *Modern Accident Investigation and Analysis* (4) by Ted S. Ferry
- *MORT Safety Assurance Systems,* (3) by William G. Johnson
- "Toward a More Universal Model of Loss Incident Causation," (2) by Robert E. McClay
- *Toward a Comprehensive Human Factors Model of Workplace Accident Causation,* (12) by David M. DeJoy

There are significant commonalities and differences in these publications. It is my intent to select from them to support a logical thought process, and add my own views.

To begin with, a hazards related incident, a HAZRIN, should have a definition. For MORT, "An accident is defined as unwanted flow of energy or environmental condition that results in adverse consequences."

In STEP, "An accident is a special class of process by which a perturbation transforms a dynamically stable activity into unintended interacting changes of states with a harmful outcome."

For McClay, "A loss incident is an unintentional, unexpected, occurrence (resulting from uncontrolled hazards) which—without any subsequent events—has the potential to produce damaging and injurious effects suddenly or, if repetitive, over a long period of time."

MORT is principally unwanted energy release based, although it also recognizes environmental conditions. McClay approaches the subject of unwanted energy release and introduces "mass" as an element:

> It can be deduced and has also been empirically shown, that the actual damaging and injurious effects are produced by the release, transformation or misapplication of energy. Since mass and energy are interconvertable, the release, transfer, or misapplication of mass should also be seen as having the potential to produce these same adverse effects.

In *Accident Prevention Manual* (15), Dr. Michael Zabetakis of the University of Pittsburgh wrote: "Most accidents are actually caused by the unplanned or unwanted release of energy . . . or of hazardous materials. . . ."

It's my observation that unplanned or unwanted releases of energy and of hazardous materials are the fundamental causal factors for hazard related occurrences.

Hazards are defined as the potential for harm or damage to people, property, or the environment. Hazards include the characteristics of things and the actions or inactions of people. Building on the foregoing, this definition is now proposed:

A hazards related incident, a HAZRIN, is an unplanned, unexpected process deriving from the realization of an uncontrolled hazard or hazards, which could be an unwanted flow of energy or an unwanted release of hazardous materials, that is likely to result in adverse consequences.

A HAZRIN causation model, then, focuses on hazards that present the potential for unwanted energy release and unwanted release of hazardous materials.

Dr. William Haddon, Jr. (16, 17) was the first director of the National Highway Safety Bureau. He is credited with having developed the energy release concept. His thinking was that an unwanted energy release can be harmful and that a systematic approach to limiting such a possibility should be undertaken. That is the framework of MORT—Management Oversight and Risk Tree.

Having knowledge of Haddon's energy release concepts is necessary in professional safety practice. Application of those concepts, *extended to include unwanted release of hazardous materials,* requires a basic review of the content of hazards management practice, of how hazards-related incidents are viewed as to their causal factors, and of what the substance of hazards management proposals ought to be.

A HAZRIN causation model that establishes unwanted energy releases and unwanted releases of hazardous materials as the *fundamental causal factors* requires adoption of a point of emphasis at the very beginning of things, rather than on the eventual behavior of the employee who may be said to have acted unsafely.

In thinking about the nature of hazard related incidents that derive from an unwanted release of energy or of hazardous materials, Benner (13, 14), Ferry (4), Johnson (3), and McClay (2) are valuable resources. For Benner:

> An accident occurs in connection with an activity involving certain interrelated elements. These activities are conducted in the presence of conditions of vulnerability (interrelated with each other or the activity) which conditions must exist for an accident to be possible. An accident begins when one of the elements engaged in the activity is overtaxed beyond its ability to recover from the overload and cannot resume functioning within the limits of its capability again in the continuity of the activity. (14)

Also from Benner:

> The accident process can be described in terms of specific interacting actors [author's note: actors may be things or people], each acting in a separate and spatial logical relationship. By breaking down the events seen as the accident into increasingly more definitive sub-events, the understanding of the phenomenon increases with each successive breakdown.

Benner emphasizes multiple events proceeding on a time scale, displayed much like a musical score, having interrelationships and culminating in an outcome. Events in the sequence may occur on a single track or impact on other events on parallel tracks. Events sequences are to be plotted on parallel lines along a time scale, in accord with their place in the sequence.

Investigating Accidents With STEP (13) by Hendrick and Benner, "centers on an investigative system called Sequentially Timed Events Plotting procedures. . . . This methodology is built on systems safety technology and the Management Oversight and Risk Tree." (13)

In several of Benner's writings, appropriate recognition is given to MORT as a foundational thought process. In *MORT Safety Assurance Systems* (3), Johnson gives recognition to Benner's research. For both, charting systems are used to graphically depict events contributing to a hazards related incident from the beginning to the end of the process.

McClay's approach to causation is somewhat different but also has some similarity to Benner's postulations and the bases on which MORT is founded. McClay indicates:

> . . . not all factors which contribute to a loss incident occur at the same time just before the incident occurs . . . causal factors can be placed into two groups with respect to the temporal nature of their occurrence . . . Causal factors which exist or occur within the same specific time frame and location as the loss incident (called proximate factors). . . . and causal factors which do not exist or occur within the same specific time frame and general location as the loss incident (called distal factors). (2)

McClay's premise, as I interpret it, is that the "loss incident" occurs over time and that there may be a variety of causal factors involved. Thus there is some continuity in what McClay, Benner, and Johnson have written.

Johnson, Benner, and Ferry stress the complexity of HAZRINS. Ferry wrote: "It has been found that simple mishaps tend to be complex in terms of many causal factors with lengthy sequences of errors and changes leading to the various events. This makes it essential for the investigator to have a system, a methodology for breaking down the entire sequence of events into individual events with supporting information." (14)

Benner says: "It is the accident process which is complex and requires a detailed comprehension before it can be understood adequately to control the inherent safety problems." (13)

According to MORT: "Accidents are usually multifactorial and develop through relatively lengthy sequences of changes and errors. Even in a relatively well-controlled work environment, the most serious events involve numerous error and change sequences, in series and parallel." (3)

For the prevention of these complex and multifactorial sequences that may result in adverse results, an "intervention" is proposed by many authors. I suggest that the "intervention," to be effective, has to impact on beginnings, knowing that getting that done will not be easily accomplished. Beginnings means an organization's culture, management commitment and involvement, and design and engineering decisions.

An organization's culture determines the probability of success of its hazards management endeavors. It is not possible to have an effective hazards management program unless senior management displays by what it does that hazards management is a subject to be taken seriously, a subject to be given equal consideration with other organizational goals.

Management commitment, its policies and organization, is shown as an element in most causation models. One could argue that management commitment and involvement is not an element on a par with other elements but rather the foundation, reflection, and extension of the organization's culture from which all hazard prevention and control activities derive.

A principal goal of the safety professional should be to influence the organization's culture and its management commitment. Undoubtedly that won't always be an easy task. But in considering a causation model, it has to be understood that prevention of hazards related incidents is best accomplished where the organization's culture and the management commitment includes appropriate requirements for hazards management.

An organization's culture and its management commitment should stand as separate and sequential items in a causation model.

First evidence of the organization's culture and management's commitment is displayed through its design and engineering decisions. Where hazards are given the required consideration in the design and engineering processes, and in the redesign and reengineering processes, a foundation is established that gives a good probability of avoiding hazards related occurrences. Design and engineering also deserves a distinct place in a causation model.

And just what observations can be made from all this? If safety professionals choose to examine their causation models, which are the basis of the advice they give, this thought process is proposed for consideration, as a beginning:

1. Safety practice is accomplished through the anticipation, identification, and evaluation of hazards, and the giving of advice to avoid, eliminate, or control those hazards so as to prevent or mitigate harm or damage to people, property, or the environment. Application of that advice should attain a state for which the risks are judged to be acceptable.

2. Professional safety practice requires that advice given be based on a sound hazards related incident causation model so that the application of that advice results in avoiding, eliminating, and controlling hazards.

3. A hazards related incident, a HAZRIN, is an unplanned, unexpected process deriving from the realization of an uncontrolled hazard or hazards, which could be an unwanted flow of energy or an unwanted release of hazardous materials, that is likely to result in adverse consequences.

4. Hazards related incidents, even the ordinary and frequent, are complex in respect to the multiple and interacting causal factors that may contribute to them, sequentially or in parallel.

5. Unwanted releases of energy and unwanted releases of hazardous materials are the fundamental causal factors for hazards related occurrences.

6. All events deriving from things or from the actions or inactions of people that contribute to HAZRINS are causal factors.

7. For safety, health, and the environment, the greatest strides forward will be made in the design and engineering processes, which include the design and engineering of facilities, processes,

work stations, and tools and, very importantly, the design of work methods, the design of the procedures for accomplishing the work.

8. A HAZRIN causation model should:

- recognize that an organization's culture determines the probability of success of hazards management endeavors and is the primary influence in a causation model

- give separate recognition to management commitment as the sine qua non in hazards management, as the foundation, reflection, and extension of the organization's culture

- be anticipatory and focus on causal factors that present the potential for unwanted energy release and unwanted release of hazardous materials

- move the emphasis of safety practice to origins rather than on results, and

- give design and engineering considerations distinct status, following an organization's culture and management commitment, with importance placed on avoiding unwanted releases of energy and of hazardous materials, and on ergonomics/ human factors engineering

Professional safety practice requires the development and application of a sound causation model for hazards-related incidents. Is it not a necessity that the advice given in professional safety practice, if applied by the decision makers, actually be causal factors based and actually serve to attain a state for which the risks are judged to be acceptable? Many causation models have been published and they are greatly divergent. Their application would lead to markedly different conclusions, given a particular set of circumstances. They can't all be right.

REFERENCES

1. Ludwig Benner, Jr. "Rating Accident Models and Investigative Methodologies." *Journal of Safety Research* 16, no. 3 (Fall 1985).

2. Robert E. McClay. "Toward a More Universal Model of Loss Incident Causation." *Professional Safety*, January and February 1989.

3. William G. Johnson. *MORT Safety Assurance Systems*. New York: Marcel Dekker, 1980.

4. Ted S. Ferry. *Modern Accident Investigation and Analysis*. New York: John Wiley & Sons, 1981.

5. Ludwig Benner, Jr. "Accident Investigations: Multilinear Sequencing Methods." *Journal of Safety Research,* June 1975.

6. H. W. Heinrich. *Industrial Accident Prevention*. New York: McGraw-Hill, 1950.

7. Dan Petersen. *Techniques of Safety Management*. Goshen, N.Y.: Aloray, 1989.

8. P. L. Clemens and R. R. Mohr. *Risk Management and Aerospace Research*. September 1990.

9. William E. Tarrants. ed. *The Measurement of Safety Performance*. New York: Garland Publishing, 1980.

10. Mark Paradies. "Root Cause Analysis and Human Factors." *Human Factors Society Bulletin,* August 1991.

11. Mary Walton. *The Deming Management Method*. New York: Putnam Publishing Company, 1986

12. David M. DeJoy. "Toward a Comprehensive Human Factors Model of Workplace Accident Causation." *Professional Safety,* May 1990.

13. Kingsley Hendrick and Ludwig Benner, Jr. *Investigating Accidents with STEP.* New York: Marcel Dekker, 1987.

14. Ludwig Benner, Jr. "5 Accident Perceptions: Their Implications For Accident Investigators." *Hazard Prevention,* September–October 1980.

15. Dr. Michael Zabetakis. *Accident Prevention Manual*. Washington, D.C.: MSHA, 1975.

16. William J. Haddon Jr. *Preventive Medicine: The Prevention of Accidents*. Boston: Little, Brown,1966

17. William J. Haddon, Jr. "On the Escape of Tigers: An Ecological Note." *Technology Review,* May 1970.

Chapter 12

Comments on Hazards

Hazards have been defined as the potential for harm or damage to people, property, or the environment. Hazards include the characteristics of things and the actions or inactions of people.

I also stated that safety practice is accomplished through:

- the anticipation, identification, and evaluation of hazards
- the giving of advice to avoid, eliminate, or control those hazards, to attain a state for which the risks are judged to be acceptable

In the appendix to *Improving Risk Communication,* titled "Risk: A Guide to Controversy," Baruch Fischhoff wrote: "By definition, all risk controversies concern the risks associated with some hazard. . . . The term "hazard" is used to describe any activity or technology that produces risk." (1)

Fischhoff properly relates hazards to risks. In understanding hazards and their significance in professional safety practice, recognition must be given to the fundamental premise that *all risks to which safety practice applies derive from hazards—there are no exceptions.*

Whatever the safety program element—management involvement, safety in the design process, employee training, hazards communication, incident investigation, use of personal protective equipment, behavior modification, et cetera—its fundamental purpose is the avoidance, elimination, or control of hazards.

Hazards should be considered in the broad context of the definition given, and every element of safety programming proposed by safety professionals should serve to avoid, eliminate, or control the aspects of the "activity" and the "technology" that present the potential for harm or damage.

Whether the concern of the safety professional is occupational safety and health, product safety, environmental affairs, fire protection, transportation safety, or any other safety-related practice, the generic base is hazards. In every one of those fields of endeavor, all activities should be similarly directed to encompass both the possible actions or inactions of people and the characteristics of properties, equipment, machinery, or materials that present the potential for harm or damage.

Hazards and risks are not synonymous, though they are used interchangeably in the literature by some authors, and with great confusion. Hazards and risks may also be equated in the literature with exposures and perils. Unfortunately, the literature on hazards, risks, exposures, and perils can be baffling. It is difficult to extract clear and precise meanings of terms from it.

In the insurance-related literature, peril is a frequently found term and may be used synonymously with hazard, or risk, or exposure. Perils insured against are commonly considered to be fires, explosions, falling aircraft, windstorms, floods, automobile accidents, embezzlements, et cetera. Perils are incidents that occur when unwanted energy or hazardous materials are released, and a hazard is realized.

Safety professionals will not be understood in their discussions of hazards and exposures and risks until they have established clear meanings of those terms. Separate definitions of each of those terms are required if communication with those to whom counsel is given is to be successful.

As Grimaldi and Simonds state in *Safety Management:* "Unless there is common understanding about the meaning of terms, it is clear that there can not be a universal effort to fulfill the objective they define." (2)

In the definition of hazards given throughout this book (repeated at the beginning of this chapter), potential is the key word. If a hazard is the potential for harm or damage, and if the hazard is not avoided, eliminated, or controlled, the potential will be realized.

Two considerations are necessary in determining whether a hazard exists. Do the characteristics of the thing or the actions or inactions of people present the potential for harm or damage? And can people, property, or the environment be harmed or damaged, should the potential of the hazard be realized? Determining that the hazard can be realized is the only measure of probability necessary in identifying the existence of a hazard. Hazard identification—which means identifying the potential for harm or damage—should be recognized as a separate and distinct step preceding hazard analysis and risk determination.

To complete a hazard analysis after a hazard has been identified and evaluated—that is, to measure the possible consequences should a hazard be realized—exposure must be assessed. That assessment would be a determination of the extent of harm or damage that could occur to people, property, and aspects of the environment in a particular setting.

Risk has been defined as a measure of the probability and severity of adverse effects. That definition was taken from Lowrance in *Of Acceptable Risk: Science and the Determination of Safety* (3).

My thinking, then, goes like this:

- A hazard is the potential for harm or damage
- If the potential is realized, the result is an incident, accident, event—or a peril
- That which can be harmed or damaged is the exposure
- Risk is a measure of the probability of the hazard being realized and the severity of harm or damage to that which is exposed

Clear definitions of hazards, exposures, and risks can obviously be developed in easily understood terms that are distinct and separate from each other. Such definitions would serve well in effective communications with decision makers. It is also neccessary for safety professionals who are to participate in risk assessments to have a clear understanding of those terms.

Many methods are available to identify hazards—historical data, codes and standards, the observations of learned people, and the use of analytical methods from the simpler "what if" system to the more complex fault tree analysis. And the literature, of which there is a great deal, speaks extensively of those methods.

After a review of much of the literature and reflecting on my own experiences, I believe that there are two significant thought processes, one built on the other, for which knowledge is a requirement in professional safety practice: first Haddon's *unwanted energy release concept,* extended by Zabetakis to include the unwanted release of hazardous materials; then the concepts on which the *Management Oversight and Risk Tree—MORT* has been developed.

I encourage safety professionals to give particular attention to the energy release concept, extended to include the unwanted release of hazardous materials, and to MORT. They would do well to research the literature on these subjects with the view that they are fundamental: in understanding hazards; to the purposes and contents of the elements of successful safety programs; and to their role of giving advice to avoid, eliminate or control "any activity or technology that produces risk".

In the *Loss Rate Concept In Safety Engineering,* R. L. Browning wrote:

> Work requires the expenditure of energy, in fact, energy is measured by the work it is capable of performing. It follows that the capability to cause—the key element in the search for valid loss exposures—will be an inventory of potentially destructive energy. (4)

Browning's statements are thought-provoking. Certainly, being oriented to causal factors is a fundamental of safety practice. And it seems that "potentially destructive energy," the release of which occurs in many undesirable incidents, is directly related to the definition of hazards previously given. The question is whether safety professionals can really do their job of giving advice toward preventing or mitigating harm or damage to people, property, or the environment without an extensive base of causal data that includes an "inventory of potentially destructive energy."

Dr. William Haddon, the first director of the National Highway Safety Bureau, is credited by several authors as the originator of the *energy release theory* (5, 6). Its concept is that *unwanted* transfers of energy can be harmful (and wasteful) and that a systematic approach to limiting their possibility should be taken.

Grimaldi and Simonds in *Safety Management* (2) recognized his work. Dr. Roger L. Brauer in *Safety and Health for Engineers* (7) commented on the energy release theory and listed the ten strategies for preventing or minimizing adverse results as outlined by Dr. Haddon. In *MORT Safety Assurance Systems,* William G. Johnson also listed Dr. Haddon's ten energy management strategies and made these comments concerning them:

> Haddon systemized a set of 10 energy management strategies in a progressive order, which can be used in various combinations. . . . Haddon points out that the larger the energy, the earlier in the strategy list should control be sought. This author would add, the larger the energy, the greater the need for redundant, successive strategies and barriers. . . . The systematic review of available strategies and the creation of optimum mixes to reduce harm has not been customary in safety. . . . The application of Haddon's concepts was tested in early trials of MORT. (8)

As a result of the testing and early trials, Haddon's strategy was introduced into the MORT analysis and system operation with an extended hierarchy of thirteen preventive measures.

Ted Ferry, in *Safety and Health Management Planning* (9), wrote about energy transfers and barriers and listed "13 strategies for systematic energy control" as cited by the Department of Energy.

Although Haddon stated in "On the Escape of Tigers: An Ecologic Note," (6) that "the concern here is the reduction of damage produced by energy transfer," he also said that "the type of categorization here is similar to those useful for dealing systematically with other environmental problems and their ecology." All hazards are not encompassed by the unwanted energy release concept, examples being the potential of asphyxiation from entering a confined space filled with inert gas, or inhalation of asbestos fibers.

Dr. Michael Zabetakis, in *Accident Prevention Manual,* stated clearly what was needed to establish a complete generic causal base: "Most accidents are caused by the unplanned or unwanted release of energy . . . or of hazardous materials" (10)

It is my opinion that the hazards presented by unwanted releases of energy or unwanted releases of hazardous materials are the basic causal factors for all hazards related occurrences.

In the ideal safety management system, every activity would be causal factor oriented, which fundamentally means being hazards oriented. Each phase of the system would have as its purpose the avoidance, elimination, or control of unwanted energy releases and unwanted releases of hazardous materials. For all hazards, the potential for harm or damage can be realized only when there is an unwanted energy release or an unwanted release of hazardous materials, either of an "activity" or of a "technological" nature, or both.

For the first of the two hazards management systems I recommend as being important to professional safety practice—Haddon's unwanted energy release concept, extended by Zabetakis to include unwanted releases of hazardous materials—a composite follows of the several sources cited, along with my own extensions. Its purpose is to provide a reference base in thinking about hazards and hazards management, when offering advice both in the design process and for the elimination and control of existing hazards.

In its use, safety professionals should consciously think about both sides of the hazard spectrum—"activity" and "technology" in determining what advice might be given about hazards. As rather simplistic examples, back injuries and falls presumed to arise out of activities continue to be prominent in occupational injury statistics; characteristics of chemicals to which employees may be exposed are a matter of technology.

A GENERIC THOUGHT PROCESS
FOR HAZARD AVOIDANCE, ELIMINATION,
OR CONTROL

This outline combines the concepts of unwanted energy release and unwanted releases of hazardous materials that are believed to be basic causal factors for hazards related occurrences. But it should be understood that some of the details in the outline will pertain to both unwanted energy releases and unwanted releases of hazardous materials in some instances, or to only one of the possibilities in others.

1. Avoid introduction of the hazard: prevent the marshalling of energy or of hazardous materials
   - avoid producing or manufacturing the energy or the hazardous material
   - substitute a safer substance for a more hazardous one
   - use material handling equipment rather than manual means
   - don't elevate persons or objects

2. Limit the amount of energy or hazardous material
   - seek ways to reduce actual or potential input
   - use the minimum energy or material for the task (voltage, chemicals, fuel storage, heights)
   - consider smaller weights in material handling
   - store hazardous materials in smaller containers
   - control vehicle operation in selected areas
   - remove unneeded objects from overhead surfaces

3. Substitute, using the less hazardous
   - replace hazardous chemicals with the less hazardous
   - replace hazardous operations with less hazardous operations
   - use designs needing less maintenance
   - use designs that are easier to maintain, considering human factors

4. Prevent unwanted energy or hazardous material buildup
   - provide appropriate signals and controls
   - use regulators, governors, and limit controls
   - provide the required redundancy
   - control accumulation of dusts, vapors, mists, et cetera
   - assure the necessary training
   - minimize storage to prevent excessive energy or hazardous material buildup

- reduce operating speed (processes, equipment, vehicles)
- develop a culture requiring adherence to prescribed practice

5. Prevent unwanted energy or hazardous material release
   - design containment vessels, structures, elevators, material handling equipment to appropriate safety factors
   - consider the unexpected in the design process, to include avoiding the wrong input
   - protect stored energy and hazardous material from possible shock
   - provide fail-safe interlocks on equipment, doors, valves
   - install railings on elevations
   - provide nonslip working surfaces
   - control traffic to avoid collisions

6. Slow down the release of energy or hazardous material
   - provide safety and bleed-off valves
   - reduce the burning rate (using an inhibitor)
   - control speed (vehicles, equipment, people)
   - reduce road grade
   - provide error-forgiving road margins

7. Separate in space or time, or both, the release of energy or hazardous materials from that which is exposed to harm exposed to harm
   - install electrical lines out of reach
   - arrange remote controls for hazardous operations
   - eliminate two-way traffic
   - separate vehicle from pedestrian traffic
   - provide warning systems and time delays
   - initiate procedures that give people time to evacuate in the event of an unwanted energy or hazardous material release

8. Interpose barriers to protect the people, property, or the environment exposed to an unwanted energy or hazardous material release

- insulation on electrical wiring
- guards on machines, enclosures, fences
- shock absorbers
- personal protective equipment
- directed venting
- walls and shields
- noise controls
- safety nets

9. Modify the shock concentrating surfaces
   - padding low overheads
   - rounded corners
   - ergonomically designed tools

10. Make that to be protected more resistant to the release of energy or hazardous material
    - earthquake and fire resistant structures
    - damage resistant materials
    - acclimatization to exposures
    - physical conditioning of personnel

11. Limit the damage after the energy release or hazardous material release has occurred
    - shut off the flow of energy or hazardous material
    - initiate rescue operations
    - provide emergency medical treatment
    - stop traffic
    - put out the fire
    - isolate hazardous areas
    - et cetera

12. Rehabilitate
    - people
    - property
    - environment

I also recommend that safety professionals develop an understanding of the system safety concepts on which the Management Oversight and Risk Tree is based and the thought process it promotes.

While MORT is unwanted energy release and environmental condition oriented, those terms should not lead one to conclude that its thought process is not applicable to all hazards.

What follows has been taken from the MORT User's Manual. It is not paraphrased since the verbiage speaks so well for itself.

> The MORT logic diagram is an idealized *safety system model* based upon the fault tree method of *system safety analysis. System analysis* is a directed process for the orderly acquisition of specific information pertinent to a given system. Its purpose is to provide the basis for informed management decision. . . .

> Within the MORT system, an incident is defined as Barrier-Control inadequate or failure without consequence. An accident is defined as unwanted flow of energy or environmental condition that results in adverse consequences. [Author's note: an unnecessary distinction, as the following paragraph shows.]

> MORT suggests that an accident is usually multi-factorial in nature. It occurs because of lack of adequate barriers and/or controls upon the unwanted energy transfer associated with the incident. It is usually preceded by initiating sequences of planning errors and operational errors that produce failures to adjust to changes in human factors or environmental factors. The failure to adjust satisfactorily leads directly to unsafe conditions and unsafe acts that arise out of the risk associated with that activity. The unsafe conditions and unsafe acts, in turn, provoke the unwanted energy. . . .

> MORT is designed to investigate accidents and incidents and to evaluate safety programs for potential accident/incident situations. Two of the many basic MORT concepts are the analysis of change and the evaluation of the adequacy of energy barriers relative to persons or objects in the energy channel. (11)

Early in the application of MORT, users will be determining whether there could be a potentially harmful energy flow or environmental condition, whether barriers and controls were less than adequate, and whether vulnerable people or property were exposed.

MORT is based on the fault tree method of system safety analysis. But its logic diagram does not require statistical entries and computations for probability. MORT is presented as an incident investigation methodology and as a base for safety program evaluation.

Its thought process could also serve the safety professional well who participates in design concept discussions. MORT's foundation in unwanted energy release and environmental condition concepts is sound and its use leads to a good understanding of hazards, exposures, and risk. A first approach to MORT could be intimidating; its logic diagram is impressive and daunting, but it need not be. Thousands have successfully completed MORT seminars.

Energy sources are listed in several texts, two of which are William G. Johnson's *MORT Safety Assurance Systems* (8) and Ted Ferry's *Safety and Health Management* (9). Their major captions are listed here:

Corrosive	Kinetic/linear	Pressure—volume
Electrical	Kinetic/rotational	Radiation
Explosive	Mass, gravity	Thermal
Flammables	Nuclear	Toxic

It is not the intent in this essay to explore in detail the technology side of hazards for fear that an imbalance would be created.

An overemphasis on the "technological" aspects of hazards could inappropriately result in the "activity" aspects of hazards being subordinated to them in the proposed thought process. In taking the position that a greater attention should be given to eliminating hazards through improved design, I propose that such consideration include both the "activity" and the "technological" side of hazards. This means giving greater attention in the design process than has been typical to the work methods being prescribed, or those that can be expected, or intended customer use of a product or its possible misuse, or the possibilities of public behavior, as well as to the characteristics of things.

What I propose gives a strong emphasis to design and engineering as the initial hazards avoidance, elimination, and control method for both the "activity" and the "technology" aspects of hazards. Surely there's nothing new in that idea; earlier safety literature was based on that premise.

With the emergence of ergonomics as a more significant segment of safety and health program management, one beneficial result will be an examination of and awareness of the fallacy of much of what we have accepted for incident causation, and a greater interest in delving into the hazards that are actually the root causes of those incidents.

Safety professionals who have undertaken such an introspection because of knowledge gained from applied ergonomics will be going back to the drawing board, giving particular attention in design concept discussions to the first two items in the outline of "A Generic Thought Process For Hazard Avoidance, Elimination, Or Control"—avoid introduction of the hazard: prevent the marshalling of energy or of hazardous materials; and limit the amount of energy or hazardous material. There is a role for safety professionals in giving counsel in the design process to avoid hazards, which is fulfilled by only a few.

Understanding hazards, identifying hazards, and analyzing the consequences of hazards that are realized are the first steps in determining risk. I would like to repeat the quotation from Baruch Fischhoff's *Risk: A Guide to Controversy* (1) that appeared earlier in this paper: "By definition, all risk controversies concern the risks associated with some hazard . . . the term "hazard" is used to describe any activity or technology that produces risk." (1)

We need to know more about risk.

REFERENCES

1. *Improving Risk Communication.* Washington, D.C.: National Academy Press, 1989.
2. John V. Grimaldi and Rollin H. Simonds. *Safety Management.* Homewood, Ill.: Irwin, 1989.
3. William W. Lowrance. *Of Acceptable Risk: Science and the Determination of Risk.* Los Altos, Calif.: William Kaufmann, 1976.

4. R. L. Browning. *The Loss Rate Concept in Safety Engineering.* New York: Marcel Dekker, 1980.
5. William J. Haddon, Jr. *The Prevention of Accidents.* Preventive Medicine, Boston: Little, Brown, 1966.
6. William J. Haddon, Jr. "On the Escape of Tigers: An Ecological Note." *Technology Review,* May 1970.
7. Roger L. Brauer. *Safety and Health for Engineers.* New York: Van Nostrand Reinhold, 1990.
8. William G. Johnson. *MORT Safety Assurance Systems.* New York: Marcel Dekker, 1980.
9. Ted Ferry. *Safety and Health Management Planning.* New York: Van Nostrand Reinhold, 1990.
10. Michael Zabetakis. *Accident Prevention Manual.* Washington, D.C.: MSHA, 1975.
11. *MORT User's Manual.* For the Department of Energy. Idaho Falls: EG&G Idaho, May 1983.

Chapter 13
Comments on Risk

I have adopted Lowrance's definition of risk, taken from his book *Of Acceptable Risk: Science and the Determination of Safety,* since it applies very well to professional safety practice: "Risk is a measure of the probability and severity of adverse effects." (1)

I previously stated that all risks within the scope of safety practice derive from hazards, and that there are no exceptions. Hazards are not risks and risks are not hazards.

I believe this is the definition of hazards that is applicable to safety practice: hazards are the potential for harm or damage to people, property, or the environment. Hazards include the characteristics of things and the actions or inactions of people. That definition presumes a potential for harm or damage and that people, property, or the environment may be harmed or damaged if the potential is realized.

In producing the *measure that becomes a statement of risk,* it is necessary that determinations be made: of the existence of a hazard; of the exposure to the hazard; of the consequences should the hazard be realized as represented by a description of the extent of harm or damage to people, property, or the environment; and of the probability of the hazard being realized.

If the exercise is stopped after the completion of the hazard identification and the assessment of the possible consequences, the result is a hazard analysis. To determine risk, the probability of an occurrence must be estimated. Risk, as Lowrance very simply and aptly states, is both a measure of the probability of an occurrence and a determination of the consequences of the occurrence.

It is the role of the safety professional to anticipate, identify, and evaluate hazards and to give advice on the avoidance, elimination, or control of those hazards, to attain a state for which the risks are judged to be acceptable. In fulfilling that role, two distinct aspects of risk must be considered: avoiding, eliminating, or reducing the *probability* of a hazard being realized; and minimizing the *severity* of adverse results if the hazard is realized.

As safety practice evolves, the required attention will be given to the avoidance of hazards in design and engineering processes. As more safety professionals are successful in establishing themselves as consultants in that process, they will become more familiar with concepts of risk and how risk reduction is effectively achieved.

Along the way, questions will arise concerning what the content of professional safety practice ought to be and how it is best applied in giving advice that actually attains a significant reduction of risk. That requires taking into consideration the perceptions of risk a management staff might have, the culture of an organization with regard to hazards management, its risk tolerance, and of how priorities are set.

Those questions are avoided when all energies are required to be devoted to compliance—with laws, codes, standards, and regulations. When so doing, little thought may be given to how much the significant risks related to employment, products, properties, or the environment are actually being reduced as compliance is attained.

Indeed, it's understood that some safety practitioners are directed by their managements to apply all of their efforts toward obtaining compliance. Surely being in compliance is a laudable goal. Unfortunately, being in compliance—meeting the standards—usually means achieving a minimally accepted level of hazards management. It should not be assumed that actions taken to be in compliance with laws, codes, standards, and regulations address an organization's principal risks or that doing so, by itself, will attain effective hazards management.

A safety director tells the story of convincing his management to spend $1,000,000 to comply with OSHA standards. Later he was asked what impact the expenditures would have on the types of employee injuries and illnesses that had been experienced. Injury and illness records were available for thirty years. Not one of the reported incidents was related to the expenditures for OSHA compliance. Yes, risks of employee injury and illness were reduced through the expenditures to comply with OSHA standards, but by how much?

An article relative to program effectiveness, by Frank Burg, appeared in the July 1991 issue of Professional Safety, titled "OSHA's Voluntary Protection Program: A Safety Management Approach for the 90s": This is an excerpt:

> Dr. Karl U. Smith from the University of Wisconsin among others spearheaded arguments that enforcement of safety and health standards, though necessary, would not be reflected in injury and illness reductions because inspections are just one portion of hazard awareness and control. In 1971, I participated in an OSHA-funded project with the state of Wisconsin, which attempted to correlate findings of state inspections with causes of injuries as reported to worker's compensation. We found no significant relationship. Our recommendations to OSHA in the early 1970s were largely ignored. (2)

But now assume that the manager of your location, or the one for which you are a consultant, tells you to be certain that you have done all that you can do to:

- Keep management out of jail
- Assure that management doesn't have to face the news media because of a catastrophic or embarrassing incident related to your function, which includes safety, health, and the environment
- Help management achieve a highly effective and efficient hazards management program

Those instructions represent a typical and mixed bag with respect to risk. They compel the safety professional to do some serious thinking and balancing concerning priorities, benefits to be obtained, and implementation cost for each hazards management proposal made. In helping keep management out of jail, advice would be given to attain "compliance."

But an obligation exists for the safety professional, in giving advice on compliance, to put the related expenditures, benefits, and risks in a proper context with those risks that relate to catastrophic or embarrassing incidents, and to those risks that require attention to achieve a highly effective and efficient hazards management program.

Managers have every right to ask safety professionals, "Why do you want me to do what you are proposing, and what are the possible results if I do or don't follow your recommendations?"

But a successful communication with management personnel on risk is not possible until an understanding has been reached on the meaning of the term. That's important. Risk is a term with far too many meanings.

On any given day, managements may hear the term in a variety of contexts—in contemplating a financial venture, in considering the possible increase in the price of a commodity, in receiving a phone call from a stock broker who proposes a risk management program. It's also important that safety professionals appreciate the culture of an organization, its fears and logic concerning risk, and its tolerance for risk. Managers are risk takers; so are safety professionals.

It is not possible to attain a risk-free environment, even in the most desirable situations. *If every proposal or recommendation made by a safety professional was favorably acted on, there would still be a residual risk, even though remote, unless a product, operation, material, premises, et cetera was wholly eliminated.* Setting a goal to achieve a zero risk environment may seem laudable, but it requires chasing after a myth.

Even if safety professionals are engaged in an enterprise where resources are unlimited—a theoretical ideal that is never attained—every risk identified can not be immediately eliminated or controlled. And as a part of the management team, the safety professional shares in the risk taking when practical scheduling decisions are made to eliminate, reduce, or control risks over time, and when a management decision is made to accept a risk that is considered tolerable.

Safety professionals, then, must be capable of distinguishing the important from the less important in giving counsel toward attaining that state for which the risks are judged to be acceptable.

And the concept of acceptable risk must be addressed here. Safety was defined as a state for which the risks are judged to be acceptable. That definition implies a determination of risk, and a judgment of the acceptability of risk. Immediately, upon use of the term acceptable, these questions arise: acceptable to whom, acceptable in whose opinion?

But consider this. Every safety professional who writes a recommendation to eliminate or control a hazard makes a risk acceptability decision. It cannot be presumed that complying with the recommendation achieves zero risk. No thing or activity is risk free.

Also, in the practical world, all risks will not be eliminated. Risk acceptance decisions are made in determining which risks are and are not to be given attention. Even in deciding which risks are to be given priority consideration, an acceptability decision is made, if only for the short term.

A few safety practitioners make too much of the process of making risk acceptability decisions. They create an atmosphere of being above that sort of thing, of super-righteousness. Perhaps they do believe that if all that they proposed was favorably acted upon, a risk free environment would be achieved. Safety professionals know that is not possible. Those safety professionals who are a part of the management team will share in risk acceptance decisions.

Safety professionals must acquire knowledge of risk determination concepts to give validity to the proposals they make to reduce risk. Since there are always resource and scheduling and time constraints, the advice given should ideally include:

- risk categorization and priorities
- possible alternatives
- expected effectiveness of each alternative in risk reduction
- remediation costs

Yet risk reduction decisions will not always be made on such logically presented data.

In considering risk estimation concepts, safety professionals should be cautiously critical of the applicability of precise risk measurement systems, particularly in dealing with the perceptions the public may have of risk. Richard F. Griffiths, in *Dealing with Risk,* wrote: "The applicability in risk assessment and acceptability is that for low-frequency events the probability estimate is not based on a large number of trials and the public evaluation may well be more conditioned on how bad the outcome might be, with little regard for arguments as to how likely it is." (3)

In this one paragraph, Griffiths introduces two important subjects: that probability estimates used for low-frequency, high-consequence incidents may be questionable; and that public concerns may reflect only perceptions of severity of outcomes.

At a symposium on risk sponsored by the National Safety Council and ILSI Risk Science Institute, the theme for which was "Regulating Risk: The Science and Politics of Risk", one speaker expressed the view that public risk concerns would best be considered as public outrage. Even though at times it may be believed that the public outrage is illogical, it will often be a significant factor in risk decision making. All risk reduction measures may not be based entirely on risk assessment logic. And it may be that in dealing with public outrage, or employee concerns, or perceived risks, facts will not be convincing.

It is not unusual for risk decision making to include elements of fear, dread, and perceived risks of employees, the immediate community, a larger public, and management personnel.

Dr. Vincent T. Covello of Columbia University expressed the view at the symposium that communicating on risk with the public is a separate skill and that much of what we do is wrong. He cautioned about the many "word traps" we may get into. In a film clip shown, for example the speaker used the term industrial hygienist. A subsequent study showed that over 60 percent of the public thought that the term industrial hygienist referred to the janitor.

At this symposium, fault tree analysis was mentioned many times. A composite of those citations, and of several readings, is that fault tree analysis is great for qualitative analyses but it is difficult to apply as a quantitative risk assessment method because of the necessity to speculate too often on probability data, and because of cost. One speaker even referred to fault tree analysis as metaphysics, requiring a great faith to believe in it. That goes a little too far.

The speakers made it evident that great variations exist in the methods government agencies use to assess risks. That had been previously recognized and it was announced that a committee is to develop a consensus on the principles and practices to be applied in risk assessments made by government agencies. It's possible that more is expected from risk assessment, which is largely judgmental, than science can deliver.

Dr. Richard B. Belzer, an economist in the U.S. Government's Office of Management and Budget, commented on part of the Fiscal 1992 Budget of the United States Government, "IX.C. Reforming Regulation and Managing Risk Reduction Sensibly". If the premises of this important document hold, safety professionals may find themselves in more discussions of risk assessment, of relative risks, and of cost benefit analysis. Brief excerpts from the document are given here:

- benefits of Federal regulations [are to] exceed the costs
- cost effectiveness "per premature death prevented" resulting from government regulations varies greatly
- the range of those costs shows clearly that society's resources for reducing risks are being poorly allocated
- the standards for determining Government spending and regulatory program priorities are haphazard, leading to great disparities in benefits in relation to cost
- the Administration is advancing a new risk management initiative that should enable decision makers to reallocate scarce resources to produce both lower risks and lower costs

This is an example of the concerns expressed: "Most significant OSHA safety standards are . . . cost effective. But, EPA regulations and OSHA health standards often entail very large costs for every premature death prevented." It's of interest that a definition of risk given in the document is compatible with Lowrance's definition.

This document has already had an effect in Federal regulatory agencies. Consider its possible impact on safety professionals and their thinking about risks. It presumes agreement on these factors: methods to assess risks; a means of determining risks that are acceptable; what standard is to be put into effect to attain the acceptable risk; a system to determine the costs to comply with the standard; and a way to quantify the benefits that would be attained through the proposed risk reduction measures. And the goal is that resources to reduce risks would be sensibly allocated.

Although this logic process is also applicable to a safety professional's role in giving advice on risk, I wonder what the expectations might be on risk quantification.

In using Lowrance's definition of risk, it is not required that a quantification be made to achieve a precise numerical result, expressed in dollars, as is proposed by several authors. Such a computation may derive from a formula, common in the literature, through which probability is multiplied by consequence to obtain the expected loss per unit of time or activity. This is a typical formula:

Risk = Probability x Consequence

>Where probability is the event frequency per
>unit of time or unit of activity, and
>consequence is loss or cost per event

Similar formulas abound, intended to produce a finite, numerical quantification of risk. Their use presumes a knowledge of incident probability far more extensive and precise than seems to exist.

It is my belief that—except when statistical concepts can be applied to the known, such as the face markings and shapes of dice, or when empirical probability evidence has been produced—all of what are called quantitative risk analyses are really qualitative risk analyses, since so many subjective judgments are necessary in applying what are otherwise sound statistical quantification systems.

Vernon L. Grose, in *Managing Risk: Systematic Loss Prevention for Executives,* (4) appropriately cautioned on several occasions against an overreliance on probability numbers in determining risk when the numbers may not be sound. A reasoned skepticism would serve safety professionals well concerning the validity of the numbers used in what appear to be precise determinations of both the probability of an occurrence and assessment of its consequences, both of which are necessary in estimating risk.

Both probability and consequence numbers may be greatly overstated or understated. Grose may have exaggerated only slightly when he wrote:

> Because desire overrode reality, the unfortunate gamesmanship that has evolved supposedly to upgrade reliability and safety has come to be called "numerology," defined in the dictionary as follows:
>
>> A system of occultism (hidden, secret, or beyond human understanding) involving divinations (the practice of trying to foretell the future by mysterious means) by numbers.

Grose indicates that when there is a "demand for numbers" and an "inventing" to fill the need, the numbers eventually "appear in print" and "there is no way to recognize their illegitimacy". This caution is offered—be wary of the numerologists.

Grose is not alone in expressing concern about the validity of risk quantification systems. *Improved Risk Communications,* a text prepared under the guidance of the Committee on Risk Perception and Communication as a project of the National Research Council, states: "Analysts are prone to overlook the ways human errors or deliberate human interventions can effect technological systems. . . .The need to quantify risks as an aid to decision making creates special difficulties because the choice of which numerical measures to use depends on values and not only science." (5) Emphasis is given to the phrase "the choice of which numerical measures to use depends on values and not only on science."

Quoting from Griffiths again in *Dealing with Risk:* "If one compares the experts' best estimates of the consequences of an accident with the historical record it is often found that the estimates are greatly in excess of the consequences actually manifested." (3)

Having had experience over many years with estimates of possible property damage loss developed by fire protection engineers after establishing reasonable-worst-case hazard and exposure scenarios, my observation is opposite Griffiths' view. Damage estimates were often considerably short of what subsequently occurred.

Very recently, I was presented with risk data to which a scoring system was to be applied. Risk priority ratings were to be computed for management attention. An incident with a probability of 1 per 100 plant operating years for which devastating results were calculated, including many fatalities, was given a lower priority than a single disabling back injury with a probability of 1 per plant operating year. A skepticism concerning numerical risk scoring systems is in order. Indeed, be wary of numerical risk systems, and the numerologists.

Nevertheless, if safety professionals are to understand risk and include risk determinations in the counsel they give, there has to be a way to establish reasonable and accepted bases to obtain agreed-upon understandings about probabilities and consequences. Much of their work will require the gathering of opinions of knowledgeable people in developing hazard/exposure scenarios and in making assumptions about occurrence probabilities.

Matrices, of which there are many in the literature, through which risks are measured and categorized on an informed judgment basis as to both probability and consequence and as to relative importance, seem to have great credibility.

Some will argue that a statement describing risk must be monetary. Nevertheless, the purposes of Lowrance's definition of risk, and the requirements of risk categorization matrices, can be met without attempting to be so precise.

This is an example of a risk estimate that does not include money. Assume that the population of the United States is at 250,000,000 and that the number of fatalities resulting from automobile usage annually is 46,000. Then, the risk of fatality, *the consequence,* was approximately at a *probability* of 1 per 5400 of population per year.

In the use of hazard analysis and risk assessment matrices, judgments of probability and consequence will often be made by experienced people with less precise data than is given in the preceding example. And such systems can be made to work. They should be considered more art than science.

REFERENCES

1. William W. Lowrance. *Of Acceptable Risk: Science and the Determination of Safety.* Los Altos, Calif.: William Kaufmann, 1976.
2. Frank Burg. "OSHA's Voluntary Protection Program: A Safety Management Approach for the 90s." *Professional Safety,* July 1991.
3. Richard F. Griffiths. ed. *Dealing With Risk.* Manchester, U.K.: Manchester University Press, 1981.
4. Vernon L. Gross. *Managing Risk: Systematic Loss Prevention for Executives.* Englewood Cliffs, N.J.: Prentice Hall, 1987.
5. *Improved Risk Communications.* Washington, D.C.; National Academy Press, 1989.

Chapter 14

On Hazard Analysis and Risk Assessment

All hazards do not present equal potential for harm or damage. All risks do not have equal probability of occurrence, nor will adverse effects from those occurrences be equal.

A safety professional has an implicit responsibly to give the advice that, if applied by the decision makers:

- permits an efficient use of resources to avoid, eliminate, or control hazards, *on a priority basis;* and
- results in attaining a state for which the risks are judged to be acceptable.

Such advice can not be given when multiple hazards are to be considered unless:

- hazards are analyzed and categorized as to their potential and the possible severity of consequences, should the potential be realized;
- estimates are given of the probability of occurrences;
- the costs are determined of alternate proposals to reduce risks; and
- judgments are made of the risk reduction benefits to be attained.

To be more precise, professional safety practice requires that hazards be analyzed, that risks be assessed, and that a ranking system be applied when giving advice on multiple hazards.

Why would so much be made of hazard analysis and risk assessment as a requirement of professional safety practice in dealing with multiple hazards? This, typically, is what happens. A hazard is identified and it is assumed that injury or illness or damage can result if the hazard is realized. Or a hazard may be considered a violation of a standard or a regulation. Almost automatically a recommendation to eliminate or control the hazard is then presented to decision makers by the safety practitioner. Little thought may be given in the process to the significance of the hazard in relation to other hazards, to the priority it ought to have, or to the possible effectiveness of the measures proposed in achieving risk reduction.

Later, it may be determined that actions taken didn't work, and other measures will be proposed, ad infinitum, perhaps also ineffective. In the meantime, more significant risks are ignored. That doesn't speak well of real accomplishment or of effective resource utilization.

In the real world, the safety professional has to recognize that:

- resources will always be limited
- some risks are more significant than others
- the greatest good to society, to employees, to employers, and to users of products is attained if resources are applied to obtain risk reduction effectively and economically.

Safety professionals must seekout and give priority to those hazards that, if realized, will result in the severest harm or damage.

Professional safety practice requires an occasional review of time expenditures and a reorienting of activities in relation to actual needs and opportunities to avoid, eliminate, or reduce hazards. That would result in devoting appropriate time to those risks that have been determined to present greater probability and severity of adverse consequences.

Thus, safety professionals must have within their endeavors a separate and distinct activity that attempts to seek out hazards that particularly present severe injury or damage potential.

In support of that position, I refer to what is known as Pareto's law. Pareto observed from analyses of monetary patterns that the significant items in a group will usually constitute a relatively small portion of the total. Those in financial fields often refer to Pareto's findings as the 80–20 rule, with 20 percent of the statistical body representing 80 percent of the total impact.

I have personally observed over the years, but not through scientific study, that Pareto's principle generally applies to employee injuries and illnesses, fires, auto incidents, product liability incidents, and pollution incidents.

One relative study of substance pertaining to employee injuries and illnesses, made by Employer's of Wausau and known as "The Vital Few" (1), included several hundred thousand cases. It indicates that 86 percent of total workers compensation injuries represent only 6 percent of the total cost; 14 percent of total injuries represent 94 percent of total costs; and 2 percent of total injuries (a part of the 14 percent) represent 70 percent of total costs.

In actual comparisons, the experiences of individual companies will not fit the given distribution precisely. But many safety directors with rather large statistical bases have also observed that the principle generally applies.

Obviously, the 14 percent of total injuries representing 94 percent of the total costs include the most severe injuries. Now assume that resources are limited and that a safety professional shares the responsibility of having resources effectively applied.

In giving advice to eliminate or control hazards, societal, employee, and employer priorities would be considered. From all viewpoints, it is obvious that hazards presenting the potential for the most severe harm or damage, those for which consequences are most costly, should be given the highest priority consideration.

Whatever the field of endeavor of the safety professional, Pareto's law applies, and it prompts some interesting questions about whether we spend far too much time on the insignificant.

Only a few authors have proposed that the identification and evaluation of severity potential deserve a special place in a hazards management program. Such a study would encompass hazard analysis and risk assessment and the giving of priority consideration to those hazards with the greatest potential for harm or damage.

In *Techniques of Safety Management,* under "Severity versus Frequency," Dan Petersen wrote: "If we study the mass data, we can readily see that the types of accidents resulting in temporary total disabilities are different from the types of accidents resulting in permanent partial disabilities or in permanent total disabilities or fatalities." (2) Petersen then gave his view of "The Ten Basic Principles of Safety," the second being: "We can predict that certain circumstances will produce severe injuries. These circumstances can be identified and controlled."

Petersen's text indicates that for these types of situations severe injuries are fairly predictable: unusual, nonroutine work; nonproduction activities; sources of high energy; certain construction situations; many lifting situations; repetitive motion situations; psychological stress situations; and exposure to toxic chemicals.

In *Profitable Risk Control,* William W. Allison spoke of the need to address severity potential:

> A basic problem has been the need of a method to enable each facility to identify those severe risks which can result in loss of life, limb, material resources and profitability in that specific facility or in a new operation. An easy method to do so is described in this chapter. It is based on the logic and facts of the transfer of energy or toxics, the similarities of dissimilars, and learning from others' experience. (3)

To identify and eliminate or control hazards that have severe injury, illness, or damage potential, the implementation of a system specifically established to seek those hazards is required.

Understandably, decisions will be made to act or not to act on certain hazards for a variety of reasons. Regardless, for that decision making, the safety professional has a responsibility to provide hazard analyses and risk assessments in a language that management understands. This implies that a safety professional has adopted a hazard analysis and risk assessment thought process and that there has been an effective communication with decision makers on those methodologies and, especially, on the meanings of terms to be used in their application.

In "Comments On Risk" (Chapter 13), I cautioned about the

improbability of having accurate and precise numbers representing probability and consequences in making hazard analyses and risk assessments. I suggested that safety professionals be wary of numerologists, and of risk scoring systems.

That position is a reflection of my personal experiences, through which I developed an awareness of the folly of making elaborate computations based on ill-supported assumptions. While doing so may create the appearance of being scientific, that's an unprofessional position to take when the work being done is based on many subjective judgments.

It was the common practice of fire protection engineers on my staff to compute property damage and business interruption loss estimates. A reasonable-worst-case hazard and exposure scenario, a modeling of an event, would be written, including assumptions about hazards and about hazards being realized, where on the property in question the incident would most likely occur, the value of the facilities and equipment in that area, and the monetary value of damage to property if there was a damaging incident. Values used were provided by the client, and their accuracy was often unbelievably questionable.

If chemicals were involved that upon release could produce a vapor cloud, the procedure was to make assumptions about the quantity of material released, its characteristics, the point of release, the shape of the cloud, wind direction and speed, the barometric pressure, et cetera.

Following a long-used concept, the energy potential of the released material was converted into an equivalency of TNT. Computations would be made and circles drawn indicating blast overpressures in pounds per square inch. Damage within, say, the 10 p.s.i. circle would be considered complete. Reduced damage levels would apply as the distance from the blast point increased.

And infinite and excruciating detail was computed for a variety of blast overpressure circles, forgetting the subjectivity of the assumptions originally made. That was done even though it was understood that if one variable was changed slightly in the beginning, the outcome would be dramatically different.

Client personnel would participate in developing hazard and exposure scenarios. They understood the impact of making changes in the variables for which assumptions had been made and that loss estimates at best would be largely subjective.

I made the decision that, for the purposes of producing such loss estimates, it would be assumed that the value of the salvage within the 3 p.s.i. circle would equal the value of the damage outside that circle. That greatly limited the amount of calculation and avoided the appearance of being overly scientific. Thus, the property damage loss estimate would be the equivalent of 100 percent of the value within the 3 p.s.i. circle. That became known, initially in jest, as the Manuele Theory of Loss Estimating. And it was applied by many.

Recently, Charles A. Pacella, Vice President and National Coordinator for Property Services at M & M Protection Consultants, completed a comparative study that included a loss estimating system used extensively in the United States, two such systems commonly used in Europe, and the Manuele Theory of Loss Estimating. He wrote: "The conclusion from this comparison is that the simplified method requires the least time and effort, has the fewest assumptions, and yields credible, realistic results."

Why should that history be relevant? First, it indicates that in situations where subjective judgments prevail at the outset, little is gained in attempting laborious computations as the hazard analysis proceeds. Also, while making those computations may give the appearance of the endeavor being scientific, in actuality those computations are not of real value.

Results of those studies represented one-half of a risk assessment. They began with hazard identification and, presuming that the hazard was realized, considered that which was exposed to determine the possible consequences. *What was produced was a hazard analysis, always assuming the reasonable-worst-case.*

Incident probability estimates were not made, nor expected. Probability data was not available. Nor, since the staff always worked with the reasonable-worst-case, were scenarios categorized or ranked in relation to others. In the advice given to clients, emphasis was placed on the hazard and exposure scenario and the estimate of the severity of consequences being plausible.

Recommendations were given to eliminate or control hazards, accompanied by the approximate costs to take the actions proposed. Frequently alternate risk reduction measures were a part of those proposals.

Making a hazard analysis precedes and is necessary to making a risk assessment. A hazard analysis concludes with an estimate of the severity of the consequences of a hazard being realized. A hazard analysis does not require that the probability of an incident occurring be determined.

As Lowrance stated in *Of Acceptable Risk: Science and the Determination of Safety:* "Risk is a measure of the probability and severity of adverse effects." (4)

Estimating the probability of the hazard and exposure scenario being realized is also necessary in making a risk assessment. Determining both probability and consequences is a requisite when giving advice on multiple hazards.

But it's of interest that estimating incident probability will not be necessary in fulfilling the hazard analysis requirements of OSHA's rule for *Process Safety Management of Highly Hazardous Chemicals* (5). It has been said that the standard could apply to as many as 50,000 places of employment. A great many of those locations are not within what would be considered chemical companies.

Hazards analyses required by OSHA is a subject of itself that deserves a review by safety professionals, even though they may not be immediately effected this new rule. OSHA's standard requires that:

> The employer shall perform an initial hazard analysis (hazard evaluation) on processes covered by this standard. The process hazard analysis shall be appropriate to the complexity of the process and shall identify, evaluate, and control the hazards involved in the process. . . .

> The employer shall use one or more of the following methodologies that are appropriate to determine and evaluate the hazards of the process being analyzed. . . . What-If; Checklist; What-If/Checklist; Hazard and Operability Study (HAZOP); Failure Modes and Effect Analysis (FMEA); Fault Tree Analysis; or An appropriate equivalent methodology. . . .

The hazard analysis shall address: The hazards of the process; The identification of any previous incident which had a likely potential for catastrophic consequences in the workplace; Engineering and administrative controls applicable to the hazards and their interrelationships . . . ; Consequences of failure of engineering and administrative controls; Facility citing; Human factors; and, A qualitative evaluation of a range of the possible safety and health effects of failure of controls on employees in the workplace.

Under the caption "Pre-startup safety review," the employer is required to provide a process hazard analysis, among other things, for new facilities and for significant modifications.

In no place in the standard is there a mention of probability. This appears in the preamble to the standard.

"OSHA has modified the paragraph [on consequence analysis] to indicate that it did not intend employers to conduct probabilistic risk assessments to satisfy the requirement to perform a consequence analysis."

Yet safety professionals will have to consider incident probability in ranking hazards and in giving advice to decision makers.

Hazards analyses are to be performed by an appropriately qualified team. Also, "The employer shall establish a system to promptly address the team's findings and recommendations; assure that the recommendations are resolved in a timely manner and that the resolution is documented." This is interesting terminology. Findings are to be addressed, and necessarily resolved.

That does not indicate, as does OSHA's general duty clause, that: "the employer shall furnish to each of his employees employment and a place of employment which are free from recognized hazards that are causing or are likely to cause death or serious physical harm to his employees."

Nevertheless, the rule requires that: "The employer shall correct deficiencies in equipment that are outside acceptable limits . . . before further use or in a safe and timely manner when necessary means are taken to assure safe operation." Note the term *acceptable*. This standard could have an impact on the involvement of safety professionals in hazards analyses, far beyond the scope of the chemical industry. Some of its features should be considered as precursors of things to come.

Several models have been published that will help the safety professional develop a suitable hazard analysis/risk assessment system. In addition to those previously cited, these are some of the methodologies mentioned in the literature: Preliminary Hazard Analysis; Gross Hazard Analysis; Hazard Criticality Ranking; Catastrophe Analysis; Energy Transfer Analysis; Human Factors Review; The Hazard Totem Pole; and Double Failure Analysis. There are other hazard analysis systems. P. L. Clemens discussed twenty-five such systems (6).

For references on hazard analysis and risk assessment, these are suggested:

- "A Compendium of Hazard Identification and Evaluation Techniques for System Safety Application" (6), by P. L. Clemens
- *System Safety Engineering and Management* (7), by Harold E. Rowland and Brian Moriarty
- *Handbook of System and Product Safety* (8), by Willie Hammer
- *MORT Safety Assurance Systems* (9), by William G. Johnson
- *Safety and Health For Engineers* (10), by Roger L. Brauer
- *Managing Risk: Systematic Loss Prevention For Executives* (11), by Vernon L. Grose
- *Military Standard—System Safety Program Requirements* (MIL-STD-882B) (12)

Hazard analysis methodologies may be deductive, moving from the general to the specific. Fault tree analysis is such an example. Or they may be inductive, moving from the specific to the general, as in failure modes and effects analysis.

While the literature speaks of many hazards analyses and risk assessment techniques, I propose that each safety professional develop a matrix and a thought process that can be comfortably implemented and that suits client needs.

In considering the development of a matrix and a thought process for its application, an awareness of reality will help:

- Because of staffing and time constraints, it is not possible to know of or to analyze all hazards. And all hazards are not equal, which requires that the safety professional make subjective but learned judgments concerning what hazards to study.

- Be aware that some situations defy statistical analysis.

- Keep it simple—there's no need to complicate things.

- Hazard analysis is an art. It is not a science. And the same applies to risk assessment. It is not possible to be absolutely certain in determining either consequences or probability. A variation of a thought attributed to Descartes applies—if you can't know the truth, you ought to seek the most probable. As a practical matter, the result of the exercise should be the best possible decision, with an understanding of the aspects of relative risk severity, costs to reduce risks, available resources, and scheduling constraints.

- It is unprofessional to give the appearance of being scientific, using volumes of charts and numbers, when risk assessments are principally subjective.

- To communicate with decision makers, terms must have been defined. As a part of the development of a hazard analysis/risk assessment matrix, it is necessary that the safety professional communicate on its contents, and obtain agreement on its use. Indeed, the literature contains several methodologies. Although identical terms may be used in those systems, they may have different meanings.

- In communicating with decision makers, it would be well to understand their perceptions and tolerance of risk, and appreciate that perceived risks as well as elements of employee and public fear and dread, and client concerns, may impact on risk decisions.

- Implementing a logical hazard analysis/risk assessment model is more important than which model is chosen.

Several texts include hazard analysis/risk assessment decision matrices. All of the matrices I found have been adopted from *Military Standard—System Safety Program Requirements*, known as MIL-Std-882-B (12). Figure 14-1, reflecting the texts reviewed and MIL-Std-882-B, is my adaptation.

Hazard Analysis/Risk Assessment Decision Matrix

Severity of Consequence	Occurrence Probability				
	Frequent	Probable	Occasional	Remote	Improbable
Catastrophic					
Critical					
Marginal					
Negligible					

Risk reduction required

Requires written, time-limited waiver, endorsed by management

Operation permissible

Figure 14-1

This matrix assumes that:

- if the Severity of Consequence is deemed Catastrophic and the Occurrence Probability is Frequent, Probable, or Occasional, or

- if the severity of Consequence is deemed Critical and the Occurrence Probability is Frequent or Probable, that

  Risk Reduction Is Required

- if the Severity of Consequence is Catastrophic and the Occurrence Probability is Remote, or

- the Severity of Consequence is Critical and the Occurrence Probability is Occasional, that

  A Written Time-Limited Waiver, Endorsed By Management Is Required

- if the Severity of Consequence is judged to be Negligible, that

  Operation Is Permissible

The Operation Permissible category does not assume that action will not be taken to eliminate or further control the identified hazards.

What do the terms in this matrix mean? A safety professional must make that determination and obtain agreement on what the terms are to designate with the people with whom there is to be communication.

Designations under the heading "Severity of Consequence" would be chosen as a result of identification of a hazard, an evaluation of its potential, and an exposure determination. An example of how those terms might apply to personal injury and property damage follows:

SEVERITY OF CONSEQUENCE

	Personal Injury	Property Damage
Catastrophic	Fatality	> $10,000,000
Critical	Severe Injury	$5 to $10,000,000
Marginal	Minor Injury	$1 to 5,000,000
Negligible	No Injury	< $1,000,000

Terms in the left hand column can mean whatever is agreed upon by the safety professional and decision makers.

Similarly, agreement would be reached on the meanings of terms under the heading "Occurrence Probability." Some guide points follow.

OCCURRENCE PROBABILITY
FOR A UNIT OF TIME OR ACTIVITY

Frequent	Likely to occur repeatedly
Probable	Will occur several times
Occasional	Likely to occur sometime
Remote	Unlikely but possible to occur
Improbable	So unlikely, probability can be considered close to zero

How does one go about this hazard/risk assessment exercise? In every one of the following steps, the safety professional should be seeking the counsel of experienced personnel within the organization who are close to the work or process, as a team, and attempt to attain their consensus.

1. *Identify the hazard.*

   This first step should be thought of separately; safety professionals should conscientiously adopt a frame of thinking in all their endeavors that promotes getting to the base of causal factors, which is hazards. A determination would be made of the potential for harm or damage that arises out of the characteristics of things and the actions or inactions of people. I also propose that the hazard potential be kept separate at this point in the thought process, which prompts recognition of severity potential for itself.

2. *Describe the exposure—the people, property, or the environment that may be harmed or damaged.*

   This is still an identification activity. Its purpose is to establish and get agreement on the number of people, the particular property, and the aspects of the environment that could be harmed or damaged, should the hazard be realized. It is not easy to do. The smart safety professional will solicit help from members of the team, from knowledgeable sources. Many judgments, more than one might realize, will be made in this process.

3. *Assess the severity of consequences.*

   Several authors propose, and I agree, that a scenario be written, preferably with the concurrence of the team, which identifies the hazard and its potential, that which is exposed, and the consequences of the hazard being realized. Such a scenario cannot be written—ought not to be written—without the participation of those who would subsequently consider its importance. Learned speculations are to be made on the number of fatalities or injuries or illnesses, on the value of property damaged, and on the extent of environmental damage.

   On a subjective judgment basis, then, agreement would be reached with participants on the severity of consequences. In the Hazard Analysis/Risk Assessment Decision Matrix, four severity categories are given from which one would choose—catastrophic, critical, marginal, and negligible. At this point, a hazard analysis will have been completed.

4. *Determine the probability of the hazard being realized*

Unless empirical data is available, and that would be a rarity, the process of selecting the probability of an incident occurring will again be subjective.

Probability has to be related to intervals of some sort, such as a unit of time or activity, missions, units produced, life cycle. Probability terms given in the Matrix—frequent, probable, occasional, remote, improbable—are commonly used. Probability should be selected with participation of "the team."

5. *Conclude with a statement that addresses both the probability of an incident occurring and the expected severity of adverse results.*

Assuming that agreement had been reached on the concepts to be applicable in a risk assessment matrix, the conclusions drawn would be a simple statement categorizing probability and consequences.

All of the preceding outline pertains to individual hazard and exposure scenarios. To properly communicate with decision makers, a risk-ranking system should be adopted. In some of the literature, risk ranking systems given appear to be overly rigid, requiring usage of a fixed procedure or formula. Since much of the hazard analysis and risk assessment exercise is subjective, I suggest that safety professionals work closely with the team in drawing risk ranking conclusions.

Some of the risk assessment systems that include a factor for the cost of reducing risk within the risk assessment and priority determination methodology give consequences classed as catastrophic and critical a subordinate ranking if the cost to reduce risk is substantial. That could be highly unprofessional. Not that costs to reduce risk and the benefits to be obtained are not to be a part of the management equation. It's just that it seems best to consider those costs separately from the risk assessment.

Consequences graded as catastrophic and critical must be addressed. Most such events would have a low probability. In the decision making for those events, an organization's culture would be determinant—its values, its sense of responsibility, its concern for its employees and the public, its determination of the risk it can bear, et cetera. Aspects of an organization's culture cannot be fit into a scoring system.

If a safety professional initiates a more formal procedure for hazard analysis and risk assessment, thought should be given to how those additional measures might succeed. There has to be a realization that significant change in management practice is being proposed and that the fundamentals applicable to achieve change would apply. Too much at one time may be disruptive.

A planned effort will be necessary to convince decision makers that the safety professional's counsel on hazard analysis and risk assessment can be of value. Small steps forward, proving value, are recommended. In preparing for such an endeavor, I suggested that safety professionals:

- develop an awareness of the risk tolerance beliefs held by decision makers

- study the approach to be made to decision makers, considering history, risk tolerance, their needs, and how decision makers would conclude that what is being proposed is of value in their achieving their goals

- working with a team of knowledgeable people, obtain agreement on the benefits to be obtained from the use of additional hazard analysis/risk assessment systems, the methodologies to be used, and the meanings of terms in the Hazard Analysis/Risk Assessment Decision Matrix

- determine which risks deserve priority consideration

- select one or two higher category hazard/risk situations and write scenarios on them

- establish a logical set of proposals, in concert with the team, being able to establish that they are soundly based and can be supported with confidence, and include alternate hazards management proposals, as appropriate

- determine the costs to implement the hazards management proposals and the expected risk reduction that is expected from each

- try, assess, modify, and try again

If hazards analyses and risk assessments are not well done, questions could arise about the soundness of what safety professionals propose to avoid, eliminate, mitigate, or control hazards. It must be understood that hazard analysis is the first step in the safety process, and that the quality of all else in the safety process follows the quality of hazard analysis.

REFERENCES

1. *Pareto's Law and the "Vital Few."* Wausau, Wis.: Employers of Wausau.

2. Dan Petersen. *Techniques of Safety Management.* Goshen, N.Y.: Aloray, 1989.

3. William W. Allison. *Profitable Risk Control.* Des Plaines, Ill.: American Society of Safety Engineers, 1986.

4. William W. Lowrance. *Of Acceptable Risk: Science and the Determination of Safety.* Los Altos, Calif.: William Kaufmann, 1976.

5. *Process Safety Management of Highly Hazardous Chemicals.* OSHA Standard at 1910.119, February 1992.

6. P. L. Clemens. "A Compendium Of Hazard Identification and Evaluation Techniques for System Safety Applications." *Hazard Prevention.* March–April 1982.

7. Harold E. Rowland and Brian Moriarty. *System Safety Engineering and Management.* New York: John Wiley, 1991.

8. Willie Hammer. *Handbook of System and Product Safety.* Englewood Cliffs, N.J.: Prentice-Hall, 1972.

9. William G. Johnson. *MORT Safety Assurance Systems.* New York: Marcel Dekker, 1980.

10. Roger L. Brauer. *Safety and Health For Engineers.* New York: Van Nostrand Reinhold, 1990.

11. Vernon L. Grose. *Managing Risk: Systematic Loss Prevention for Executives.* Englewood Cliffs, N.J.: Prentice-Hall, 1987.

12. *Military Standard—System Safety Program Requirements* (MIL-STD-882-B). Washington, D.C.: Department of Defense, 1987.

Chapter 15

Successful Safety Management: A Reflection of an Organization's Culture

An organization's culture determines the probability of success of its hazards management endeavors. What the board of directors or senior management decides is acceptable for the prevention and control of hazards is a reflection of its culture.

Thus, as a derivation of its culture, management gets that hazards related incident experience that it establishes as tolerable. For personnel in the organization, "tolerable" is their interpretation of what senior management does.

An organization's culture consists of its values, beliefs, legends, rituals, mission, goals, emphases, performance measures, and sense of responsibility to its employees, to its customers, and to its community, all of which are translated into a system of expected behavior.

Articles and chapters of books have appeared with titles such as "The Hazard Control Process," "Basic Safety Programming," "Managing Safety Performance," "Management of Loss Control," "How Do You Know Your Hazard Control Program Is Effective?" But none of the authors considered the impact of an organization's culture on the safety performance attainable.

Where the prevention and control of hazards is done best, an understanding is achieved at a board of directors level or at a senior executive level that the organization is to have superior performance in avoiding harm or damage to people, property, or the environment. This culture is demonstrated through the systems in place for performance measurement, accountability, and communications; through the design and engineering decisions for its facilities, equipment, and products; and through the specifics of the safety management program.

It is not possible to have an effective hazard prevention and control program unless the senior executive staff displays by what it does that hazards management is a subject to be taken very seriously, a subject to be given equal consideration with other organizational goals.

What management does, rather than what management says, permeates the thousands of decisions that are made in creating the work environment, setting design specifications for facilities and equipment, establishing fire protection standards, responding to occupational health needs, managing environmental affairs, et cetera.

This sense of responsibility carries down from the board of directors or through the senior executive staff, whose decisions both establish and reflect the organization's culture. Those decisions fundamentally affect the probability of success for all aspects of hazards management. As examples, they will determine:

- the priority level safety will receive
- the risks to be accepted
- the design and engineering standards that establish the quality of the work environment
- the expenditures to be made to eliminate or mitigate identified hazards
- the frequency and severity of events to be considered tolerable (employee injuries and illnesses, fires, product liability incidents, auto accidents, environmental incidents)
- whether a product is to be recalled or eliminated
- the effectiveness of emergency planning

Since effective safety management is so heavily influenced by an organization's culture, it would be prudent for safety professionals to determine what values have been established, what's important, management's tolerance for risk, what's attainable, and how things get done—thereby learning of and experiencing the organization's culture. Variations of risk tolerance, particularly by industry, are great.

Understanding an organization's culture is vital to the success of the safety practitioner. It is essential for the safety professional to be perceived as a part of the management team. Failure will result if the safety practitioner ignores management goals and is perceived as not assisting the decision-makers in fulfilling their expectations.

This is not to suggest that the safety practitioner should be less than professional. If accomplishment is the safety professional's purpose, then an understanding must be attained of the priorities of managers at a given time (expansion, contraction, capital expenditure restrictions, increased competition, severity of expense control, staffing constraints) and of the organization's culture and how to work effectively within it.

A principal goal for safety professionals should be to influence the organization's culture as it pertains to safety decision making. Understandably, this goal may not be attained easily.

When giving advice on hazards management improvements, safety professionals should be asking:

- Is elimination or mitigation of the hazard, only, sufficient?
- Does the culture condone the hazards noted?
- For long-term effectiveness, is there a need for a culture change and, if yes, how it is best achieved?

Affecting a culture change doesn't get done quickly (a supertanker can't make a sharp right turn). An organization will experience the impact of the culture in place for quite some time. Significant cultural improvement or deterioration occurs only in the long term.

Because of rising costs, public embarrassment, or a number of other factors, management may decide that dramatically improved results must be attained in a rather short time. A bit of skepticism is appropriate when that occurs.

Several studies have reported on successful safety program management, although these have been no recent writings that address the significant changes in emphasis that have occurred in applied hazards management in recent years. A composite follows of the current practices in several companies in which the culture requires accomplishment through its safety endeavors, along with my personal observations.

Management direction and involvement is the sine qua non, the prime requirement for safety program effectiveness. What senior management does is interpreted by the organization as the role model to be followed. Although senior management serves as the role model, it is important to understand that success in incident prevention can only be achieved through complete involvement down through all levels of employment.

It's at the senior management level that measurable goals are established for performance expectations. Over the long term, effective safety management is demonstrated by the results achieved in relation to performance expectations.

As an example, in a 190 year old organization in which it is understood that the chief executive officer is the chief safety officer, OSHA injury and illness rates are consistently about one-tenth of the all-industry national average. This is an organization in which its people publicly profess that safety is a part of its heritage, that safety is good business, and that safety makes the company credible.

This is a brief summary of its worldwide corporate safety, health and environmental policy:

- We will comply with all laws and regulations in all manufacturing, product development, marketing, and distribution activities.
- We will routinely review our operations to upgrade beyond legal requirements.
- We will ensure each product can be made, used, handled, and disposed of safely.
- We will inform employees and the public about our products and workplace chemicals.

• We will provide leadership to communities to respond to emergencies.

That policy is believed by most employees to be a sincere statement applied by management as written. It establishes a proactive rather than a reactive posture and implies anticipatory prevention and protection and the allocation of the resources necessary for accomplishment.

Safety and health considerations, it has been stated, are equal to other business concerns. Accountability for safety performance is clearly established with line management at every level.

Companies that stress off-the-job safety are rare exceptions. This company has established that off-the-job safety is as important as on-the-job safety, recognizing that the factors that influence behavior and beliefs are the same in both environments.

In another organization that also achieves exceptionally good results, the management commitment starts with an Environment, Health, and Safety Committee consisting of five members of the Board of Directors. Three of the committee members—including the chairman—are "outside" directors. Establishment of this committee sent a strong and visible message of intentions and accountability throughout the company, and had a significant impact on its culture.

This company's published "Values" states: "We pledge to protect the environment and the health and safety of employees, the users of our products, and the communities in which we operate."

In its Annual Report for 1990, the company's long-term goals and strategies are given under the headings "Financial," 'People," "Progress," and "Environment, Health, and Safety." The latter section offers an extension of its "Values" statement: "We will maintain a leadership position in the protection of the environment, the health and safety of our employees, contractors, the users of our products, and the communities in which we operate." Comments pertaining to the stated environment, health, and safety goal are included in the 1990 Annual Report for each major operating company. All this is indicative of a culture that requires commitment, involvement, and accountability.

This company has slightly over 50,000 employees. Results as measured by its records regarding the environment and the safety and health of employees are commendable. It is now in the process of obtaining agreement on improvement goals that are rarely achieved. As a matter of its culture, no fatality is ever considered acceptable for any reason.

Establishment of accountability for performance in safety, health, and environmental affairs is one of the principal measures of management commitment. If an accountability system that permeates the entire organization has not been established, management commitment is questionable. Here are two real-life indicators of accountability in practice:

- A plant manager, speaking at a conference, said that the first items discussed in his annual performance review were his achievements in relation to previously established goals for employee injuries and illnesses, environmental occurrences, and fires. Meeting or not meeting those goals had a bearing on his salary. He was very much informed about incidents that had occurred and his involvement was readily apparent. He could quote his fire losses to the mil, in relation to plant values.

- A company became displeased with its employee injury, motor vehicle, and product liability incident experience. Its senior executives made arrangements to visit the facility of another organization, known to have a superior incident record. When discussions commenced, visitors were surprised that the meeting was run by the manager of the host location. It became obvious that the safety program was the manager's program and that he considered himself accountable for it.

He spoke in depth of his personal involvement in capital expenditure considerations for hazard control, of his requirements for the safety and health professional staff, of the system in place through which he maintained accountability, and of his expectations of the staff immediately reporting to him. During the plant tour, which he led, he commented extensively on the specifics of hazard control measures in the facility, displaying his personal involvement.

An accountability system that works requires the establishment of agreed-upon performance goals with financial recognition if the goals are or are not met.

In a previous example it was said that the chief executive officer in a certain company was the chief safety officer. At its locations, managers are to function as chief safety officers, and are to be the role models. Managers cannot delegate or abdicate this responsibility.

A top-quality safety organization and staff are a requisite for accomplishing intentions as defined by the culture. An organization's personnel will "read" the importance given to safety by management through its appraisals of the qualifications of the safety staff and their reporting place in the management structure. If the safety director's position is treated as insignificant, management instructs the organization that safety is insignificant. There is no one magic reporting structure for the safety function, except that the senior safety executive is not very far from the top in companies where results are superior.

In one such organization, the Vice President for Safety, Health and the Environment reports to the Senior Vice President for Human Resources and Corporate Plans who reports to the Chief Executive Officer. In another company, the Vice President for Environmental Affairs and Safety reports to the Executive Vice President who reports to the Chairman. In this company, a Committee of the Board of Directors provides oversight and requires accountability for environmental, health, and safety matters.

In both of these organizations, safety (in a very broad sense), health, and environmental affairs have been brought together under a single management. That is the trend. An awareness has developed that the basic sciences of safety, health, and the environment overlap considerably and that greater management effectiveness can be attained under a single direction.

Equally important is the need for effective communication among the professionals involved in each hazards related function. A good case can be made for a unified management that includes all hazards related professionals.

As important as the reporting relationship is the quality of the safety staff and how it is perceived. At all levels, safety personnel serve in a staff role. They are expected to earn recognition and respect and establish their capabilities, thereby being sought by decision makers for their views. They are considered to be a part of management, and have ready access to senior decision makers.

Professional requirements for safety personnel in terms of education, experience, accomplishment, and executive ability in those organizations that incorporate effective safety programs within their cultures have been moved up a few notches in recent years.

Communication and information systems pertain to all of the elements of safety management for which comment is made in this essay, but particularly to the accountability system. Management at all levels as well as employees are informed of the hazards of operations and of what is expected in relation to them. They are made aware of progress relative to established goals. Reports to senior management on results are given serious treatment as a part of the accountability system.

Design and engineering considerations for new and altered processes, equipment, and facilities become the first, outwardly evident indications of the organization's culture concerning hazard prevention and control. Its culture is seen in the quality of its buildings and properties, in the standards applied to create a safe place in which to work, in its requirements for fire protection and life safety, in the attention given to user safety in product development, in its standards for vehicles, in the quality of its environmental controls, et cetera.

Where hazards are given the required consideration in the design and engineering processes, a foundation is established that gives good probability to favorable incident avoidance. Also, the potentially large expenses of retrofitting are avoided.

As a reflection of the best of organizational cultures, design and engineering specifications are established to go beyond legal requirements, are intended within good judgment to avoid unacceptable risk, and consider both the characteristics of things and the possible actions or inactions of people.

This is a subject that has not been given sufficient attention by safety professionals. The subject of design and engineering considerations is included in this essay, but it would not typically appear in outlines of safety management systems, nor would it be included as a subject in safety audits.

Yet design and engineering decisions are primary in determining the eventual risk level, and they are most often made without input from safety professionals. Thus safety professionals are typically confronted with the workplace, equipment, and products as givens, with thousands of design and engineering decisions affecting safety having been made without their counsel.

As the safety profession evolves, as a better understanding is developed of the phenomena of incident causation, and as ergonomics as an emerging art and science impacts further on the practice of safety professionals, they will be giving greater attention to design decisions. It has been a rewarding experience when safety professionals have been sought for their counsel in concept and design decision making. There is both need and opportunity here for advice to be given by safety professionals.

Preventive maintenance, which in the most sophisticated operations includes a predictive element, obviously impacts greatly on hazard prevention and control. The quality of maintenance sends messages to the entire staff, informing them of the reality of the company's intent to keep or not to keep physical hazards to a minimum. Visit a location where the culture imposes good safety practice and you will get a "feel" for the quality of preventive maintenance from the appearance of the exterior premises.

That isn't necessarily an absolute indicator, but the opposite is almost always true: if the outside is shabby, safety maintenance will more than likely be inadequate. In the best operations, cleanliness is truly a virtue, maintenance schedules are adhered to, and personnel are encouraged to seek elimination of hazards.

This is an opposite and real picture. Assume that a safety professional is making an audit of the quality of hazard prevention and control. The maintenance superintendent displays an elaborate computer-based maintenance program of which he is very proud.

During the plant tour, a multitude of hazardous conditions are observed. A supervisor is asked why work orders aren't being sent to the maintenance department to have those conditions corrected. And the response is: "We don't do that anymore. Safety work orders are the last priority for the maintenance department." Later it is determined that a great number of safety-related work orders are over six months old. But the maintenance program, on paper, was supposed to prevent that sort of thing from happening.

Consider what message a situation of that sort delivers. If the staff is to believe that hazard prevention and control is to be taken seriously, management must maintain a safe environment and continuously demonstrate its commitment to do so.

Safety committees will exist at several levels. Where they are programmed to achieve, they serve as a means of communicating that hazard prevention and control is important within the organization's culture, provide for participation by a large part of the company, can be structured to allow greater employee involvement and upward communication, are well organized, have clearly understood purposes, and their recommendations are seriously considered and are brought to a conclusion at appropriate management levels.

Although articles have been written questioning the value of safety committees, they are made to work where superior results are expected and achieved. At senior management levels, their purposes may be accomplished when the management committee also serves as a safety committee and safety is an early item on the agendas for their meetings.

Surely if safety committees are not effective, their existence can become a detriment to good hazards management. Where safety committees are effective, the following practices are *not* typical: meetings are scheduled for a particular time and place and frequently canceled; a majority of the management personnel who were to attend find other things to do that they consider to be more important; early in the meeting, the management members who did attend are called away on more urgent matters; members of committees get bogged down in discussions on insignificant detail leaving no time to make proposals of importance; recommendations made by the committee are ignored.

Supervisory participation and accountability become vitally important once it is understood that the hazard prevention job gets done through the line organization. Staff groups, safety professionals, and safety committees exist to support line personnel.

Supervisory participation will directly reflect the location manager's interpretation of the organization's culture, what the manager believes is expected, and what the manager understands to be the actual performance measures.

Supervisors will do what they perceive to be important to their superiors. If their superiors convey, by what they do, that hazard prevention and control is relatively unimportant, be assured that supervisors will so respond. If supervisors are not made aware of their responsibility for the prevention and control of hazards and held accountable, failure certainly will result.

Expectations of supervisors, by their superiors and by society in general, have unfortunately become complex and difficult to attain, which means that supervisors must have a sound support structure to be successful. That support structure begins with the location manager and the staff immediately subordinate to the manager. It includes depth of training, a good communication system on hazards, up and down, and the resources of qualified safety professionals as consultants.

Training encompasses senior management, supervisors, and line employees. Unfortunately, safety training is often much talked and written about but poorly done. In companies with superior safety records, training is serious business. In the best situations, training needs are anticipated along with plans for new or altered facilities or the planned use of new materials, and consider the changing aspects of the work force and the continuing stream of regulatory requirements.

Senior management, in the model companies, is "well trained." It all starts here. All levels of management are responsible and accountable for incident prevention. They are aware of the risks of their businesses and acquire the necessary knowledge of hazard prevention and control needs. They cannot be role models and provide the necessary leadership if they are not schooled in how the hazard prevention and control job is to be done.

But safety is integrated among the functions of those with operations responsibility. And accomplishments are achieved through the supervisory staff and employees. An appropriate measure of the organization's culture as respects hazard prevention and control is the extent of involvement and the beliefs and practices of employees and their supervisors.

Safety training has to be particularly well done at these two levels. It must be well planned, continuous, and measured for results. It's not enough just to conduct a supervisor's safety training program. Supervisors have to believe that the content of the training program is what management expects them to apply, as a matter of the organization's culture, and that it serves real knowledge and skill requirements. Safety training, given as a replacement for other primary hazard prevention and control measures that should be taken, will be viewed for what it is.

Several authors have cautioned that employers should not consider safety training as the primary method for preventing workplace incidents, and that premise is sound. Rather, the first course of action should be to determine whether the hazards can be eliminated through design and engineering measures.

Training will be less effective if known hazards are not corrected. Understandably, if those hazards continue to exist, the purposes of the training program will be questioned.

Employees cannot be expected to follow safe work practices if they have not been instructed in the proper procedures. They need to understand when they begin employment that they have entered an organization that gives high priority to safe performance. It's typical to have a very thorough indoctrination procedure, a common purpose of which is to emphasize that people are the most important element of an occupational safety and health program. As new employees pass through the indoctrination program and are assigned to a supervisor, they are very quickly able to evaluate the level of safety expected.

Too much emphasis cannot be given to the importance of the supervisor in employee training. Whatever role models the supervisor and other more experienced employees provide will be followed by the new employee. It is impossible to overemphasize the high place given to employee safety training where safety management is most successful.

But consider this situation as representative of a reality that is too prevalent. Rather early during an audit of safety management, an industrial relations director reviewed in great detail with the auditor a marvelous indoctrination and safety training program for new employees. During the course of the audit, an interview was arranged, at random, with an employee who it turned out had been in the shop for about three months. It was the intent to determine what he thought of the indoctrination and safety training program.

His response was, "What indoctrination and safety training program?" This employee had bid up to his third job, had never gone through the indoctrination and safety training program, complained that he never saw his supervisor, and didn't know how to get anyone to pay attention to gear box covers that had been removed and not replaced. Situations of this sort define the organization's culture for hazard prevention and control.

Employee involvement in many aspects of hazards management has been extended in recent years by some companies, with great success. In a broad sense, it has been the premise that extended employee involvement builds confidence and trust in the organization, develops more enthusiastic and productive employees, and supports the position that all are working together to achieve understood objectives.

A realization has developed that when employees are brought into the responsibility scheme of things, they can make substantive contributions in hazard identification, in proposing solutions, and as participants in applying those solutions. Safety and health programs obviously are more effective if employees have "bought into them." Their involvement must be meaningful and include decision-making capability.

As an example, practitioners in ergonomics tell countless stories of work practice innovations originating from first line employees. Many are easy to apply, inexpensive, and effective, and often result in greater productivity. There is an asset here, in greater employee involvement, that could be better utilized to achieve more effective hazards management.

Purchasing standards are influenced by many facets of a culture. This influence comes from design and engineering personnel, from the management decision makers, and very much from safety professionals. An understanding of corporate goals is achieved and communication is open and continuous, with a respect for the positions of all involved.

Hazards analyses processes are an integral and important element in successful hazards management. In the best cases, hazards analyses are anticipatory. They are a part of concept discussions, design and engineering decisions, and process reviews. From a risk assessment viewpoint, the proper questions are: Is there a hazard? Can its potential be realized? What is exposed to harm or damage? What will the consequence be if it does happen? How often can it happen? Hazards analyses may be completed through mechanisms as simple as checklists, something more detailed such as job hazard analyses, or in complex cases, fault tree analyses.

Whatever the mechanism, the goal is always the same: hazards are to be anticipated, identified, and evaluated, and the appropriate prevention and control measures are to be determined. A safety professional will be highly skilled in hazard analysis methods and will participate in training others in their fundamentals.

Ergonomics and human factors engineering are emerging to become more significant as elements of successful safety and health programs. A few companies now have ergonomics specialists in their headquarters or location staffs who serve as consultants to those who make workplace or product design decisions. In the literature, "ergonomics" and "human factors engineering" are becoming synonymous and are used interchangeably. One university gives courses in both, the difference being that ergonomics covers workplace design and human factors engineering extends the study to include product design.

Companies whose cultures require a continuing pursuit of improvement in safety program effectiveness have taken a leadership role in exploring the possibilities that ergonomics presents, particularly with respect to the prevention of employee injuries and illnesses. For those purposes, ergonomics can be defined as the art and science of designing the workplace to fit the worker.

At the Western Safety Conference held in Anaheim in 1991, Gerard F. Scannell, then Assistant Secretary of Labor for Occupational Safety and Health, said that it's estimated that 50 percent of OSHA reported incidents were ergonomics related. A major workers compensation insurer has concluded that about 50 percent of reported claims and about 60 percent of their attendant costs had ergonomics implications. This information is relatively new and different from causal data that safety professionals have previously considered to be sound. It requires serious introspection by safety professionals concerning the content of their practice and how they spend their time.

Recognition of ergonomics as a requirement for successful safety and health program management will have a great impact on the knowledge and skill requirements of safety professionals and on their relationships with decision makers. Ergonomics, as its significance emerges, will also promote a greater recognition of the impact of workplace design decisions on both risk reduction and productivity.

Control of occupational health hazards has been a major emphasis of OSHA since its beginning and most companies have given the subject priority attention. Occupational health professionals are now frequently a part of hazard prevention and control staffs, and expenditures to control health hazards are great. Surely, keeping occupational health hazards at an acceptable risk level is a must.

Control of environmental pollutants has been approached both as a matter of good citizenship and out of the concern over costly penalties that might be imposed by EPA in those companies where the culture compels the avoidance of damage to the environment. This is a subject that often gets greater senior management attention than other aspects of hazard prevention and control and is influenced by public perceptions and a typically aggressive press. It is common in the best management structures for those responsible for environmental affairs to have senior level credentials. And they have management support to achieve.

Safe practice standards are established, communicated, and implemented, with the subject being taken very seriously by employees and all levels of supervision. These standards become the substance of training programs and of expectations by supervisors. Development of safe practice standards more often involves some form of "employee empowerment" through which their input is sought. A safety program cannot succeed without soundly established and implemented safe practices. How well that's done is another reflection of the culture.

Inspection programs will exist at several levels where hazard prevention and control are best managed. They have many purposes, one of the most important being that they display management's determination that hazardous conditions and practices are to be identified and corrected. They also provide meaningful opportunities for participation by a cross-section of all employment levels. The most effective inspections are those in which senior executives participate. Correcting observed hazards is a demonstration of the culture. Failure to follow through, of course, gives negative messages.

Incident reporting and investigation is one of the major factors that determine how the staff "reads" what level of hazard prevention and control is really acceptable to management. Do it poorly, and poor readings are inevitable. How does it get done well? Management has to be a part of the accountability for investigations in some manner. In one company, the plant manager is expected to participate in 10 percent of incident investigations. In another worldwide company, the location manager (not the safety director) must report to headquarters within forty-eight hours on any injury resulting in lost work days.

Absolutely, there has to be a documented incident reporting and investigation procedure that encompasses incidents that have the potential for serious injury, damage, or environmental pollution as well as those occurrences that have had such results. But that's not enough. Specialized training is necessary to achieve a sophistication in incident investigation. That's recognized in what could be considered the model companies. Incidents don't occur in a given department very often, and those who investigate them have limited experience in doing so. Thus the necessary training is regularly repeated.

This is quite a challenge to safety professionals, who must have a thorough understanding of concepts of incident causation and the ability to conduct effective training programs.

For incidents that have the potential for or have resulted in serious consequences, it's common to assemble investigation teams representing the required talents.

Most importantly, results of investigations are publicized and the necessary corrective actions are implemented promptly. Quality of incident reporting and investigation tolerated is a principal measure of the accountability system, and of the culture of which the accountability system is a part.

It's very difficult to achieve effectiveness in other aspects of hazards management if corrective action is not taken to correct the causal factors for the incidents that do occur.

Recording, analysis, and use of incident data follows directly in importance to incident investigation. If the accountability system is to work, there has to be an effective incident information gathering and analysis system. Culture requirements that include a review of whether management personnel meet or do not meet agreed-upon goals rely on this system. Also, the analytical data produced is vital in determining where hazard prevention and control emphasis needs to be given. This function is well performed in relatively few places. More often, incident information gathering and analysis fits very well under the axiom "garbage in, garbage out."

Performance measurement and communication is multifaceted. Data produced by the incident recording and analysis system is, of course, a principal aspect of performance measurement. But other measures of the quality of safety performance are utilized in good safety programming. If the sole measure of performance is based on incident occurrences, it should be understood that such a system measures only those defects arising out of operations in a given time, which would not necessarily provide a measure of the quality of safety management.

Additional proactive measurement and communication systems are needed, systems that serve managerial goals and involve a cross-section of personnel. One such system is the Critical Incident Technique, the use of which grows because it seeks employee involvement. Another is the safety audit, being performed best on a "user friendly" basis, thus serving to assist managers in reaching their performance goals. Inspection programs also serve as quality of performance measurement techniques and ought to be so viewed.

Through the Critical Incident Technique, a trained observer would interview a sampling of experienced employees to obtain their comments on incidents that could have resulted in serious injury or damage but didn't, and on hazards they know exist, both technological and activity based. It would be made clear that the purpose of the endeavor is to identify hazards, especially those that may be critical, so that they may be corrected before incidents occur. It is essential that observations made by employees followed through to a proper conclusion.

Safety audits are a part of many safety management programs. They are proactive and intended to measure the quality and effectiveness of the safety processes in place and to serve as the basis of an evaluation of what is and is not working well. Safety audits are well done by but a few safety practitioners.

If the safety professional wants to be considered a part of the management team, the approach to making safety audits should be one of assisting the location manager in attaining operating goals. So, rather than seeming to be a removed and unsympathetic critic, the safety professional should take a supportive approach. Too many safety practitioners have considered their job finished when the safety audit report was completed, without any consideration of the additional help that might be needed in high-risk identification, in priority setting and in programming the corrections proposed.

Safety audits are exceptionally important: the receiver of them should view the safety professionals who completed them as partners in attempting to achieve hazard prevention and control goals. A good test might be whether the manager would invite the safety professional who made the audit to come back.

Technical information systems that serve as a resource on hazard prevention and control subjects exist in all organizations where safety expectations are very high. Extent of use of the resource is a reflection of the quality of the safety, health, and environmental affairs staff.

Medical and first-aid facilities are superior at both a corporate and at a location level where a sense of responsibility to employees permeates the culture.

Emergency and disaster planning require senior management attention and programs are quite good in some places. With great sympathy, it needs to be said that it's very difficult to put in place and maintain programs that are seldom used. A few companies have periodically tested their programs, after selecting an event that plausibly could occur and communicating the assumption of its occurrence to those who would be involved. Emergency and disaster planning programs cannot be maintained without regularly testing them. And the training requirements are considerable and continuing. Only those companies dedicated to protecting its employees and its community provide the resources necessary to establish and maintain sound emergency and disaster planning.

Compliance programs concerning government regulations, while last in this listing, are given much importance at a corporate level and attention to them percolates down through entities that are to maintain top-quality safety, health, and environmental management programs. But compliance programs do not determine operating standards. It's common in the best situations for government regulations to be considered basic standards, with actual design and operating requirements often exceeding them.

• • •

It would be unusual for a listing of the elements of successful hazards management programs not to commence with management commitment and involvement. One could argue that management commitment and involvement is not on a par with other elements but rather the foundation, reflection, and extension of the organization's culture from which all hazard prevention and control activities derive. Management involvement is absolutely required, and is the principal measure of management commitment.

In entities that have achieved outstanding safety records, the organization knows that management is held accountable, is involved, and holds subordinates responsible for their results.

If incident experience is considered to be unsatisfactory by management, safety professionals should promote, with great tact and diplomacy, the asking of the obvious but difficult questions. Has that experience resulted from an absence of commitment to hazard prevention and control? Has the adverse experience been programmed into operations, by implication?

It is impossible for superior safety performance to be attained if executive personnel do not display, by their actions, that they intend to have it. Management is what management does. What management does establishes the organization's culture. If what management does gives negative impressions, it is unlikely that a hazard prevention and control program will be successful.

Emphasis must be placed on what is done in relation to what may be written or said. People will not be fooled for very long by what is said if what is done is perceived to be substantively different. Consider this example.

In a major industrial firm, the cost for the workers compensation insurance renewal premium increased dramatically because of worsening frequency and severity of claims. A letter was sent by the C.E.O. to all operating executives stating that the cost increase was not tolerable and that a reduction was expected.

That's all. No change in accountability. No change in reporting requirements. No change in priorities. No additional resources. That organization understood its culture and had learned how to interpret such a letter. They ignored it.

Hazards management programs fail when personnel perceive that what management does gives an indication that management's interest in hazard prevention and control does not include requiring accountability for safe performance. Involvement can be demonstrated in many ways: by setting high standards for facilities and equipment, by setting high but attainable performance requirements, by regularly communicating on safety subjects, by serving as chairperson of a safety committee, by leading discussions of safety performance, by directing selected incident investigations, but most importantly, by clearly establishing and giving strong emphasis to accountability.

In recent years, more senior executives have taken a stronger and more active position on safety, health, and the environment. Where this has occurred, a significant culture change is achieved. One can well imagine that in those organizations there is a constant management commitment, a well-understood accountability and a determination to achieve good results.

REFERENCES

1. Daniel R. Denison. *Corporate Culture and Organizational Effective-ness.* New York: John Wiley & Sons, 1990.

2. Harold J. Leavitt. *Corporate Pathfinders.* Homewood, Ill.: Dow Jones-Irwin, 1986.

3. John V. Grimaldi. and Rollin H. Simonds. *Safety Management.* Homewood, Ill.: Irwin, 1989.

4. R. L. Browning. *The Loss Rate Concept in Safety Engineering.* New York: Marcel Dekker, 1980.

5. Roger L. Brauer. *Safety and Health for Engineers.* New York: Van Nostrand Reinhold, 1990.

6. Ted Ferry. *Safety and Health Management Planning.* New York: Van Nostrand Reinhold, 1990.

7. Dan Petersen. *Techniques of Safety Management.* Goshen, N.Y.: Aloray, 1989.

8. David S. Gloss. and Miriam Gayle Wardle. *Introduction to Safety Engineering.* New York: John Wiley & Sons, 1984.

9. William G. Johnson. *MORT Safety Assurance Systems.* New York: Marcel Dekker, 1980.

Chapter 16

Anticipating OSHA's Standard for Safety and Health Program Management

Within the next few years, OSHA will have promulgated a general industry Standard for Safety and Health Program Management. Safety professionals, you can so advise your senior managements. Tell them, if you like, that Fred Manuele says there is a 93.7 percent certainty of it. And you can inform them that the provisions of the standard, although subject to public review and comment, have already been written. They are included in this essay.

How would I arrive at those conclusions? That inference derives from a review of certain aspects of OSHA's history and of relative congressional legislative history.

It is the intent of the Occupational Safety and Health Act to "assure so far as possible every working man and woman in the Nation safe and healthful working conditions." Its general duty clause requires that "Each employer—shall furnish to each of his employees employment and a place of employment which are free from recognized hazards that are causing or are likely to cause death or serious physical harm to his employees [and] shall comply with occupational safety and health standards promulgated under this Act."

Procedures are set forth in the Act for the promulgation of standards, when OSHA "upon the basis of information submitted . . .

determines that a rule shall be promulgated in order to serve the objectives of this Act."

Some of the safety professionals employed by OSHA are aware that the requirements of employers and the purposes of the Act cannot be met unless well-managed safety and health programs are in place. Eventually, they surmise, a Standard for Safety and Health Program Management would have to be promulgated.

For quite some time after the enactment of OSHA, of necessity, its Standards writing was confined to specific hazards. In the early 1980s, steps were taken by OSHA that resulted in the development of a safety and health management program and gave recognition to the significance of the program by removing sites that met its requirements from OSHA's programmed inspection lists.

OSHA's Voluntary Protection Program (VPP) was adopted in 1982. Changes, not significant, were made in 1988. The Federal Register entry for July 12, 1988, through which the latest changes were accomplished, states:

> Requirements for VPP participation are based on comprehensive management systems with active employee involvement to prevent and control the potential safety and health hazards of the site. Companies which qualify generally view OSHA standards as a minimum level of safety and health performance and set their own stringent standards where necessary for effective employee protection. . . .

> OSHA has long recognized that compliance with its standards cannot of itself accomplish all the goals established by the Act. The standards, no matter how carefully conceived and properly developed, will never cover all unsafe activities and conditions.

> The purpose of the Voluntary Protection Program (VPP) is to emphasize the importance of, the improvement of, and recognize excellence in employer-provided, site-specific occupational safety and health programs. These programs are comprised of management systems for preventing or controlling occupational hazards. . . .

When employers apply for and achieve approval for participation in the VPP, they are removed from programmed inspection lists. (1)

There are three VPP programs—the Star Program, the Merit Program, and the Voluntary Protection Demonstration Program. A goal of the latter two is to qualify for the Star Program.

In the preceding quotes, the term "management systems" appears more than once. That's significant. Qualifications for the Star Program are an outline for a safety and health program management system. And OSHA literature states that employers who met the requirements of that system achieved good results.

On January 26, 1989, OSHA published *Safety and Health Program Management Guidelines* in the Federal Register (2). The background data for these Guidelines stated:

> In 1982 OSHA began to approve worksites with exemplary safety and health management programs for participation in the Voluntary Protection Program (VPP). Safety and health practices, procedures, and recordkeeping at participating worksites have been carefully monitored by OSHA. These VPP worksites generally have lost-workday case rates that range from one-fifth to one-third the rates experienced by average worksites. . . .
>
> Based upon the success of the VPP and positive experience with other safety and health program initiatives. . . . [OSHA proceeded and obtained public comments on the proposed guidelines.]

Several commentors stated that the guidelines should have been mandated and enforced as a rule.

It will be difficult to fault OSHA's *Safety and Health Program Management Guidelines,* except for emphasis or detail. They describe a sound management system and are performance-oriented.

It's reasonable to assume that OSHA, in fulfilling its responsibilities, will promulgate a Standard for Safety and Health Program Management and that the standard will resemble the issued guidelines very closely.

Following are excerpts taken verbatim from the guidelines; despite their brevity, they nevertheless retain the guidelines' substance.

EXCERPTS FROM OSHA'S
SAFETY AND HEALTH PROGRAM
MANAGEMENT GUIDELINES

Issued at 54 FR 3904, January 26, 1989

Scope and Application

"This guideline applies to all places of employment covered" by OSHA, except those covered by 29 CFR 1926, which is the construction safety standard.

Introduction

Effective management of worker safety and health protection is a decisive factor in reducing the extent and severity of work-related injuries and illnesses. . . .

Effective management addresses all . . . hazards whether or not they are regulated by government standards. . . .

OSHA urges all employers to establish and to maintain programs which meet these guidelines in a manner which addresses the specific operations and conditions of their worksites.

The Guidelines

a) General

Employers are advised and encouraged to institute and maintain in their establishments a program which provides systematic policies, procedures, and practices that are adequate to recognize and protect their employees from occupational safety and health hazards. . . .

An effective program includes procedures for the systematic identification, evaluation, and prevention or control of general workplace hazards, specific job hazards, and potential hazards which may arise from foreseeable conditions.

An effective program looks beyond specific requirements of law to address all hazards. . . .

Written guidance . . . ensure(s) clear communication of policies and priorities and consistent and fair application or rules."

[Next in the guidelines comes item (b), Major Elements, of which there are four. In a following item (c), Recommended Actions are set forth for each of the Major Elements. In these excerpts, the applicable Recommended Actions follow Major Elements directly.]

Major Element 1

Management commitment and employee involvement are complimentary. Management commitment provides the motivating force and resources . . . and applies its commitment to safety and health protection with as much vigor as to other organizational purposes. . . .

Employee involvement provides the means through which workers develop and/or express their own commitment to safety and health protection, for themselves and for their fellow workers.

Recommended Actions

State clearly a worksite policy on safe and healthful work and working conditions, so that all personnel with responsibility . . . Understand the priority of safety and health protection in relation to other organizational values. . . .

Establish and communicate a clear goal for the safety and health program and objectives for meeting that goal. . . .

Provide visible top management involvement. . . .

Provide for and encourage employee involvement. . . .

Assign and communicate responsibility . . . so that [all] know what performance is expected of them. . . .

Provide adequate authority and resources. . . .

Hold managers, supervisors and employees accountable. . . .

Review program operations at least annually to evaluate their success . . . so that deficiencies can be identified . . . [and] . . . objectives can be revised. . . .

Major Element 2

Worksite analysis . . . involves a variety of worksite examinations, to identify not only existing hazards but also conditions and operations in which changes might occur to create hazards. . . .

Effective management actively analyzes the work and worksite, to anticipate and prevent harmful occurrences.

Recommended Actions

Conduct comprehensive baseline worksite surveys for safety and health and periodic comprehensive update surveys.

Analyze planned and new facilities, processes, materials, and equipment. . . .

Perform routine job hazard analyses. . . .

Provide for regular site safety and health inspections. . . .

Provide a reliable system for employees, without fear of reprisal, to notify management personnel [of perceived hazards.] . . .

Provide for investigation of accidents. . . .

Analyze injury and illness trends. . . .

Major Element 3

Hazard prevention and control . . . Where feasible, hazards are prevented by effective design of the job site or job. Where it is not feasible to eliminate them, they are controlled to prevent unsafe and unhealthful exposure.

Recommended Actions

So that all current and potential hazards, however detected, are corrected or controlled in a timely manner, establish procedures for that purpose, using the following measures:

Engineering techniques where feasible and appropriate.

Procedures for safe work which are understood and followed . . . as a result of training, positive reinforcement, and, if necessary, enforcement through a clearly communicated disciplinary system.

Provision of personal protective equipment.

Administrative controls, such as reducing the duration of exposure.

Provide for facility and equipment maintenance so that hazardous breakdown is prevented.

Plan and prepare for emergencies, and conduct training and drills. . . .

Establish a medical program. . . .

Major Element 4

Safety and health training addresses the safety and health responsibilities of all personnel concerned . . . .

Recommended Actions

Ensure that all employees understand the hazards to which they may be exposed and how to prevent harm to themselves and others from exposure to these hazards. . . .

So that supervisors will carry out their safety and health responsibilities effectively, ensure that they understand those responsibilities and the reasons for them, including:

Analyzing the work . . . to identify . . . hazards.

Maintaining physical protection in their work areas.

Reinforcing employee training on . . . hazards [and] on needed protective measures, through continual performance feedback and, if necessary, through enforcement of safe work practices.

Ensure that managers understand their safety and health responsibilities as described under . . . . Management Commitment and Employee Involvement, so that the managers will effectively carry out those responsibilities.

• • •

Developments since the issuance of the guidelines support the idea that OSHA may consider them a model for safety and health management programs.

In 1990, OSHA issued *Ergonomics Program Management Guidelines For Meatpacking Plants* (3). They are built on the same four program elements that are contained in the Safety and Health Program Management Guidelines issued in 1989—Management Commitment and Employee Involvement, Worksite Analysis, Hazard Prevention and Control, and Safety and Health Training.

In July of 1990, OSHA issued *Proposed Rule for Process Safety Management of Highly Hazardous Chemicals* (4). Its standard was promulgated in February of 1992 (5).

In at least one way, getting it done was a bit unusual. About four months after OSHA had issued its proposed standard, following its rule-making procedures, the Clean Air Act Amendments (CAAA) became law. OSHA was required by the CAAA to promulgate, pursuant to the Occupational Safety and Health Act of 1970, a chemical process safety standard, which it was already in the process of doing.

CAAA requirements about the contents of the standard that OSHA was to promulgate were specific and somewhat detailed. *Safety professionals, those requirements represent what Congress was willing to stipulate and what the Executive branch of government endorsed.* That's significant. And OSHA met those requirements meticulously.

OSHA's Standard for Process Safety Management of Highly Hazardous Chemicals was strongly supported in principle by the chemicals industry. It is a management standard and a performance standard. A multidisciplinary approach will be required in its application. These are the major headings for its requirements:

- Employee Participation
- Process Safety Information
- Process Hazard Analysis
- Operating Procedures
- Training
- Contractors
- Pre-startup Safety Review
- Mechanical Integrity
- Hot Work Permit
- Management of Change
- Incident Investigation
- Emergency Plans and Responses
- Compliance Safety Audits
- Trade Secrets

Although some of these categories may be considered unique to chemicals, all of them fit within the four major elements of the previously issued Guidelines, with the exception of Trade Secrets. Comments are appropriate on the evolution of some of these program elements.

It's of interest that Employee Participation is the first element. The background data issued by OSHA with the promulgation of the standard clearly establishes OSHA's intent with regard to employee participation:

> Employee participation . . . provisions require that employers consult with employees and their representatives on the general development of a process safety management program, as well as on the process hazards analysis. . . .
>
> OSHA believes that employers must consult with employees and their representatives on the development and conduct of hazard assessments [OSHA's hazard analysis] and consult with employees on the development of chemical accident prevention plans [the balance of the OSHA required elements in the process safety management standard]. And, as prescribed by CAAA, OSHA is requiring that all process hazard analyses and all other information required to be developed by this standard be available to employees and their representatives.

This standard, since it sets out requirements for Process Hazard Analysis and Pre-startup Review, specifically requires hazards analyses of existing facilities, and hazards analyses of new facilities and of significant modifications. Since hazards are the generic base for the existence of safety professionals, they should be active participants in the required hazards analyses. These provisions fit closely with the Worksite Analysis requirements of the Guidelines.

Management of Change means what it says: "The employer shall establish written procedures to manage changes." This part of the standard is meant to ensure hazard identification and analysis in the change process and that all other aspects of the standard—process safety information, training, and so on—are met, in relation to the findings.

In 1991, an unpublished draft paper was prepared representing the views of some of the personnel at OSHA on what should be

contained in an ergonomics standard for general industry. It dupli-
cated much of the language of the guidelines, and also followed its
four program elements. Understandably, medical management was
given much more emphasis in that paper, because of the nature of
ergonomics injuries and illnesses.

In both the House of Representatives (6) and in the Senate (7),
bills were introduced in 1991 to revise the Occupational Safety and
Health Act of 1970. The following excerpts were taken from the
Senate bill. In the House bill, the differences are of form, not substance.

> Not later than 1 year after the effective date of this section, the
> Secretary shall promulgate final regulations concerning the
> establishment and implementation of employer safety and
> health programs under this section.

And just what would the legislation require?

> IN GENERAL—Each employer shall, in accordance, with this
> section, establish and carry out a safety and health program to
> reduce or eliminate hazards to prevent injuries and illnesses to
> employees.

These are the specific requirements of the safety and health program:

> A safety and health program established and carried out . . . shall
> be a written program that shall include—
>
> 1. methods and procedures for identifying, evaluating, and
>   documenting safety and health hazards;
> 2. methods and procedures for correcting the safety and health
>   hazards identified under paragraph (1);
> 3. methods and procedures for investigating work-related
>   fatalities, injuries and illnesses;
> 4. methods and procedures for providing occupational safety
>   and health services, including emergency response and
>   first aid procedures;
> 5. methods and procedures for employee participation in the
>   implementation of the safety and health program, including
>   participation through any safety and health committee . . .

6. methods and procedures for responding to the recommendations of the safety and health committee, where applicable;

7. methods and procedures for providing safety and health training and education to employees and to members of the safety and health committee . . .

8. the designation of a representative of the employer who has the qualifications and responsibility to identify safety and health hazards and the authority to initiate corrective action where applicable;

9. in the case of a worksite where employees of two or more employers work, procedures for each employer to protect employees at the worksite from hazards under the employer's control, including procedures to provide information on safety and health hazards to other employers and employees at the worksite; and

10. such other provisions as the Secretary requires to effectuate the purposes of the Act.

Compare the requirements of these bills with the content of OSHA's previously issued *Safety and Health Program Management Guidelines*. There are great similarities. Hearings have been held on the bills. Whatever the short-term outcome of the hearings, the ideas represented in the proposed legislation will not go away. And it seems probable that personnel at OSHA have been discussing a plan of action in relation to this proposed legislation.

Safety professionals, you can depend on it with a certainty of 93.7 percent. OSHA will promulgate a Safety and Health Program Management Standard. Its framework is contained in the *Safety and Health Program Management Guidelines* issued by OSHA in 1989.

REFERENCES

1. *OSHA'S Voluntary Protection Programs.* Federal Register, 53, July 12, 1988.

2. *OSHA's Safety and Health Program Management Guidelines.* Federal Register, 54, January 26, 1989.

3. *OSHA's Ergonomics Program Management Guidelines for Meatpack-ing Plants.* U.S. Department of Labor, 1990.

4. *OSHA's Proposed Rule for Process Safety Management of Highly Hazardous Chemicals.* Federal Register, 55, July 17, 1990.

5. *OSHA's Final Rule for Process Safety Management of Highly Hazardous Chemicals.* Federal Register, February 24, 1992.

6. H.R.3160, A Bill to Revise the Occupational Safety and Health Act of 1970, August 1, 1991.

7. S.1622, A Bill to Amend the Occupational Safety and Health Act of 1970, August 1, 1991.

Chapter 17

On Management Fads

If history is indicative, new management fads will arise in the United States not less than once in three or four years. And safety professionals will be caught up in them.

What do I mean by a management fad? A new management scheme, a panacea, usually proposed by consultants or educators, will be adopted with great enthusiasm. This quick fix will solve all management problems.

With the new management scheme in place, employees will be happier, with improved morale and loyalty. Productivity and quality will surge. Growth and margins will be better than ever. And, the consultants say, managers will embrace the new ideas with enthusiasm since their advantages will be readily apparent and the new scheme will be perceived by them as making their jobs easier.

An organization's culture consists of its values, beliefs, legends, rituals, mission, goals, emphases, performance measures, and sense of responsibility to its employees, to its customers, and to its community, all of which are translated into a system of expected behavior. Success of a hazards management program will be determined by an organization's culture, which takes years to develop.

If management does not determine the impact that the new management practices will have on the organization's culture and plan well for their integration into that culture, the new fad, the quick fix, will have a short life. And if the practice of a safety professional has a sound theoretical and practical base, it will be obvious that there will be no quick fixes in achieving what needs to be done.

Management fads last but a few years, for only as long as the executive in charge continues to have an interest in them. For each of the management fads that have surfaced in the last half of this century, there has been an initial burst of excitement, a broad adoption of the fad by many organizations, and then a quiet fading away. We do what the fads require, only to find later that what we were doing becomes unimportant. But some fads will leave their remnants behind if they contained some good.

I would say to a safety professional that it would be a good idea to be a student of management styles, to develop an awareness of the culture of the organization in which the safety professional is resident, and to learn about the substance of the management fads that arose in recent years. Also, I would say that the new management fads that will surely emerge will more than likely be variations on previously developed themes.

Employee empowerment, participative management, and Total Quality Management are currently in vogue. Each is a modification of rather elderly ideas. Their life spans will tell us whether they are or are not fads.

Obviously, managements have a never-ending passion for a panacea, an aching, a penchant for instant solutions. And that passion suggests a discomfort, an insecurity, with whatever management practices are in place. It's as though management is seeking a magical cure for perceived ills. And the consultants and educators accommodate the passion. In recent years, there has been a flurry of publications and seminars, each providing its own magical potion that will achieve quick fix solutions for managerial problems.

Management fads do some good. They generate a great deal of excitement and compel managers to an introspection about how they do their jobs. But the disadvantages of management fads may outweigh the advantages.

Unfortunately, the new management practices may be adopted in place of facing an organization's real challenges. And the staff soon recognizes the gimmickry and the absence of substance in what is being proposed. Employees who have been aboard for some time quickly recognize the shallowness and the "fraud" of it all, and pass it off as just another management quirk. Employees are not easily fooled, not for very long. I have wondered about how much management fad activity is actually fakery and quackery.

How can the new management fads effect the practice of safety? It is impossible for safety professionals not to be involved. The new management methods should be put in context with sound, professional safety practice.

It is not uncommon for a safety professional to discuss, at a meeting of peers, the dynamic new scheme being adopted by management. That scheme is to revolutionize the way business is done, from top to bottom, usually with emphasis on improved personnel relations. And the safety professional may be completely caught up in the new scheme, becoming a true believer, with unrealistic expectations.

Management fads of the second half of this century, which is a little short of my personal experience with them, covered a great breadth of ideas, some of which are still applied, at least partially. While management fads may generate much excitement, they may also detract from achieving professional goals. A safety professional's judgments should be based on sound theories and principles that would be applicable regardless of the current management fad.

As an example, while I strongly support the idea of employee empowerment, whatever it's called, it's my impression that too much is expected of it and that a few safety professionals are relying on it for stupendous results that cannot be achieved.

Employee empowerment, an adaptation of McGregor's Theory Y, is the buzz phrase of the '90s. Its basis is that employees can make greater contributions if they are given the opportunity, that employees are to have involvement and participation in making decisions about the tasks to be performed. To an extent, that has been proven.

In that sort of an environment, employee contributions in ergonomics, for instance, have been frequent and notable. But at what level are they? Has the current literature not been saying that American employees are poorly educated, that they are ill-prepared for today's work needs?

Can it be assumed that those employees, about whose capability there has been much criticism, can recognize and propose corrections for basic workplace ergonomics design problems? Regardless of the atmosphere created for employees, will they be able to provide solutions that require the skills of design engineers and ergonomics.

Are the significant decisions, impacting on safety, quality, and productivity, outside of the influence of nonmanagement and nonprofessional personnel? Is it quackery to expect employee empowerment to do what it cannot do? I refer to employee empowerment only for illustrative purposes. I would ask similar questions about other management fads.

I would like to review those major management principles and practices that I believe were the most significant in the second half of this century, significant because they had the greatest influence and the most staying power. Then I will briefly comment on the management fads that have come and gone, but not necessarily entirely gone, in the past forty years.

Before I proceed with comments on the principles and methods that I believe have had the most effect on current management practice, I would like to mention Frederick W. Taylor, author of *The Principles of Scientific Management* (1). Taylor is known in some circles as "the father of scientific management." His work was widely influential but predates the term—the second half of this century—I chose to review. Why mention him? From his work, the idea developed that management could be a science. I state with emphasis that management is more art than science.

Although the business schools were to produce managers with all the tools necessary to approach management as a science, that never happened. I will never forget reviewing management practices with a successful CEO who said that the feeling in his guts took precedence over all the numbers placed before him when making an important decision. And that statement made a lot of sense to me. It related well to my own management experiences.

I believe that much of current management emphases can be traced to two thinkers and authors—Douglas McGregor and Abraham H. Maslow. I also believe that other variations on their themes, for which safety professionals will be participants, will surface in the future.

Theory X and Theory Y concepts were originated by Douglas McGregor, who has had a major influence on current management practices and on many other proponents of forms of employee empowerment and participative management. His book *The Human Side of Enterprise* (2), published in 1960, put forth his view that the approach managers took in managing was based on assumptions they made about the people they managed. These were the Theory X and Theory Y assumptions:

Theory X Assumptions

1. People do not like work and try to avoid it.
2. People do not like work, so managers have to control, direct, coerce, and threaten employees to get them to work toward organizational goals.
3. People prefer to be directed, to avoid responsibility; they want security; they have little ambition.

Theory Y Assumptions

1. People do not naturally dislike work; work is a natural part of their lives.
2. People are internally motivated to reach objectives to which they are committed.
3. People are committed to goals to the degree that they receive personal rewards when they reach their objectives.
4. People will both seek and accept responsibility under favorable conditions.
5. People have the capacity to be innovative in solving organizational problems.
6. People are bright, but under most organizational conditions their potentials are under utilized.

If a manager believed that workers had Theory X characteristics, McGregor's view was that they would be managed with an array of strategies, more of which might be considered "hard" than "soft." Employees would then need close direction and control. It was McGregor's opinion that his Theory X category represented many beliefs about human nature that were widely held by managers.

His Theory Y characteristics included contrasting views about the nature of people. Its premise is that people can be self-actualizing, exercise intellectual capacities, and more fully utilize their capabilities—if they are given opportunities to participate in establishing goals and to commit to them, and to influence the nature of their work. Concepts of employee empowerment and participative management are based on this Theory Y.

In the introduction to a later book by McGregor titled *The Professional Manager,* Edgar H. Schein made these comments about *The Human Side of Enterprise.*

> Doug's book has been seen as a plea for an attitude, a set of values toward people, symbolized by the term Theory Y. The essence of this attitude is to trust people, to grant them power to motivate and control themselves, to believe in their capacity to integrate their own personal values with the goals of the organization. Doug believed that individual needs can and should be integrated with organizational goals. In the extreme, Theory Y has meant democratic processes in management, giving people a greater voice in the making of decisions and trusting them to contribute rationally and loyally without surrounding them with elaborate control structures. (3)

In *The Professional Manager,* McGregor built on his Theory X and Theory Y theme. He also recognized the influence of Abraham H. Maslow on his work. In *Motivation and Personality* (4), Maslow set forth "A Theory of Human Motivation." McGregor stated that he had attempted to summarize a view of the motivational nature of man associated prominently with the work of Maslow.

McGregor says this of Maslow's work:

> Its central thesis is that human needs are organized in an hierarchy, with physical needs for survival at the base. At progressively higher levels are needs for security, social interaction, and ego satisfaction. Generally speaking, when lower-level needs are

reasonably well satisfied, successfully higher levels of needs become relatively more important as motivators.

In *Motivation and Personality,* the chapter on "A Theory of Human Motivation" is an attempt by Maslow "to formulate a positive theory of motivation. . . ." Maslow described a sequence of "basic human needs . . . organized into a hierarchy of relative potency."

In order of emergence, each dependent on the satisfaction of the previous need, the Physiological Needs (hunger and thirst), were followed by the Safety Needs (security, freedom from fear), and then the Belongingness and Love Needs, which if met would allow the Esteem Needs to arise, to be followed by the Need For Self-Actualization.

Maslow set forth "The Precondition for the Basic Need Satisfaction":

> There are certain conditions that are immediate prerequisites for the basic need satisfactions. Danger to these is reacted to as if it were direct danger to basic needs themselves. Such conditions as freedom to speak, freedom to do what one wishes so long as no harm is done to others, freedom to express oneself, freedom to investigate and seek information, freedom to defend oneself, justice, fairness, honesty, orderliness in the group are examples of such preconditions for basic need satisfaction.

McGregor, and several others, further developed the idea that, when other human needs were met, the need for self-actualization would arise. McGregor, in *The Professional Manager,* wrote:

> A final value is central to my view of an appropriate managerial strategy. It has to do with a motivational characteristic which Maslow called self-actualization and which other have labeled with terms like self-realization and self-expression.

As an example of how the work of Maslow and McGregor influenced other writers on management practices, this is taken from *Goal Setting* by Charles L. Hughes: "In fact, there is one purpose which the president and his employees share with all human beings—the purpose of becoming. Becoming what? More human; that is actualizing the human potential for psychological growth." (5)

One concern I have with employee empowerment, as I am being told about it in practice, is that employees may perceive what is expected of them as being manipulative. McGregor cautioned about the "influences that give the illusion, but not the fact, of choice." In that sense, he refers to commonly expressed phrases such as "Make them feel important" and "Give them a sense of participation."

There is danger in the practice of safety when employees are given opportunities to make decisions, making them feel important and giving them a sense of participation, when the decisions are recognized as not being all that significant.

As McGregor says, when manipulation is recognized, "The resultant mistrust is likely to be long-lived and difficult to overcome."

Although McGregor's Theory X and Theory Y might seem to place a manager in one camp or another, that was never his intent. McGregor did not diminish the need for authority but he did plead for a just use of it and for individual growth possibilities. While some writers have said that McGregor's theories have been disproven, they nevertheless continue to be the basis of many management approaches taken to this day.

Without a doubt, the origins of many of the management practices now in vogue—such as employee empowerment and forms of participative management—can be traced to Maslow and McGregor.

But I'm not all that certain that the majority of the population will achieve the upper realms of self-actualization that these thinkers seemed to say was natural. Nevertheless, one can readily support the idea of giving individuals oppourtunities to develop their talents, with minimum constraints.

Mention must also be made of *Management By Objectives* (MBO) as having a broad impact on management practices in many companies. Although the specifics of MBO may not be as prominently applied as previously, its concepts are still quite prevalent. Anyone who writes an annual plan, whatever it's called, is engaged in a form of management by objectives.

Management By Objectives was the brain-child of Peter Drucker who first used the term in his *The Practice of Management* (6) which was published in 1954. George S. Odiorne recognized this origin in his book *Management by Objectives* (7), a 1965 publication.

McGregor also explored management by objectives themes. In the MBO process, managers and those reporting to them would agree on organizational goals, the assignments and responsibilities of individuals toward meeting those goals, what the individual plans of action are to be, and the performance measures to be applied.

To quote Odiorne:

> Management by objectives assumes that managerial behavior is more important than manager personality, and that this behavior should be defined in terms of results measured against established goals, rather than in terms of common goals for all managers, or common methods of managing.

Now for a brief review of some of the management fads that have come and largely gone in the last half of this century. Maslow's and McGregor's influence in many of them will be apparent.

Total Quality Management (TQM) has, in many companies, replaced Zero Defects and Quality Circles programs, both of which had relatively short life spans. TQM is a hot subject. I mention it here in a list of possible management fads, but I have great hopes that TQM is not perceived in later years as another flash that appeared, stayed awhile, and disappeared.

When TQM concepts start with Deming's premise that if you want quality, you have to design it into a product or process, as in *Out of the Crisis* (8), they have a good chance of succeeding. Its potential is immense if TQM commences with a recognition of the significance in achieving success of an organization's culture, of management commitment, and of product and process design.

TQM, if done properly, requires a significant culture change in many organizations. My observations are that several companies are approaching TQM with the management commitment, involvement, accountability systems, and resources that are necessary to achieve the required culture changes.

Management by Committee was in vogue for a while. Almost all decisions were to be made by committee, which was cumbersome. On a much more limited basis, the practice continues in some organizations. In all such arrangements, someone rises to a leadership role, whether or not that person is designated as being the leader. (I think that whatever the organizational structure—traditional reporting relationships or the now popular work "teams"—someone has to be the leader, someone has to be in charge.)

Management by Consensus had a relatively short life, since it proved unwieldy and time consuming, and often resulted in less than laudable decisions.

Centralization and Decentralization are given emphasis, each in its time. I recall being told early in my career that many companies go through a ten-year cycle of centralizing and decentralizing. There is something to the existence of such a cycle.

Quantitative Management proposed that the sole measure for management decision making was numbers. In a few places, this concept may still apply.

Pay For Performance required a structuring of the work so that rewards would be in concert with performance. Application of the idea was difficult, except at higher management levels.

Intrapreneuring, which made but a brief appearance, encouraged executives to create and control entrepreneurial projects within the organization.

Management by Walking Around would require executives to visit locations more frequently, rather than manage by progress reports. A good idea, but now seldom heard of as a management theme.

Theory Z suggested the adoption of Japanese management methods (but our American workers chose not to become identical with Japanese workers.)

Transactional Analysis would have managers become more sensitive to employees, as in McGregor's Theory Y. Managers would be less bossy and promote more employee participation. A great many companies put their managers through some sort of sensitivity training. Goals were laudable but desired results were seldom achieved. It was difficult for managers to relate the content of the seminars to the day-to-day content of their jobs. Also some managers felt that they were expected to practice psychotherapy, for which they were not qualified.

Zero Defects also had its turn, and did not last very long. As a concept, it actually wasn't sound. It has been replaced in many companies by other quality programs.

Quality Circles sounded great and were widely adopted. Small groups of employees were to have regular meetings and discuss quality problems. Their recommendations for improvement in quality and productivity were encouraged and were supposed to be acted upon. Quality circles, unfortunately, worked on the secondary end of quality problems, not on the product or process design, and their results were minimal. Management did not, as a rule, seek customer involvement and increase expenditures for research and design when they created quality circles.

As several authors have written, quality must be designed into the process. Improvements that can be made on the production floor will help but they won't solve quality problems that originate in product design and process design.

The One-Minute Manager (9) was a book by Kenneth H. Blanchard and Spencer Johnson. It was the biggest rip-off I ever experienced in buying books. Its premise was absurd. In what was proposed, there were relationships to management by objectives and to behavior modification. But goal-setting was to be accomplished in one minute, as were situations for which employee praise or reprimand were in order. Its ideas could never be put in practice in real manager-employee relationships.

Matrix Management gave people with particular skills more than one responsibility, and more than one boss. When functional and administrative reporting lines were drawn on an organizational chart, the result was a matrix. It is not easy to report to two or more bosses and the idea, though with seemingly apparent values, didn't work well.

PERT/CPM is the acronym for Program Evaluation and Review Technique/Critical Path Method. Its methodology includes the identification of all activities to be undertaken in a project, charting them in relation to their sequence and completion time, determining their relationships to each other, and identifying the sequence of activities that are critical to completion of the job, on time. This idea has had staying power, in many forms.

"Excellence" was the theme of *In Search of Excellence* (10) by Thomas J. Peters and Robert H. Waterman. Without a doubt, "Excellence" made the biggest splash of any of the management fads. "Excellence" became big business. It was the thing to achieve. Peters and Waterman were management consultants. They concluded that certain characteristics were common in the management of well-run companies with which they were familiar and that the success of those companies derived from the application of those common traits. There were eight such traits.

1. Bias for action
2. Staying close to the action
3. Autonomy and entrepreneurship
4. Productivity through people
5. Hands-on, value-driven
6. Stick to the knitting
7. Simple form, lean staff
8. Simultaneous loose-tight properties

Who could argue that those characteristics are not praiseworthy? But they were not sufficient for the companies cited in the book to continue to achieve the great results attributed to them when the book was written. Although "Excellence" was pursued by many, its thesis also lost favor when some of the companies that were supposed to be models ran into difficult times.

The Managerial Grid (11) is the title of a book by Robert R. Blake and Jane S. Mouton. Its theme is participative management. Along with the formation of teams, managers would achieve a balance between concerns for production and concerns for people. By attaining the appropriate place on the managerial grid—the balance—all would be improved: employee participation, morale, contribution, sense of worth—and productivity. This is an adaptation of McGregor's Theory Y. Its ideas seemed sound, but not easily applied.

• • •

Others would say that I omitted from the preceding list some management fads of the second half of this century. And that would be so. As an additional reference, I recommend *American Business and the Quick Fix* (12) by Michael E. McGill.

My point is that new management fads, in which safety professionals will be expected to have involvement, arise regularly, produce excitement, consume a great deal of energy, and may do some good or produce only marginal benefits. As they arise, it would be prudent for safety professionals to maintain sound hazards management principles.

REFERENCES

1. Frederick W. Taylor. *The Principles of Scientific Management.* New York: Harper & Brothers, 1947.

2. Douglas McGregor. *The Human Side of Enterprise.* New York: McGraw-Hill, 1960.

3. Douglas McGregor. *The Professional Manager.* New York: McGraw-Hill, 1967.

4. Abraham H. Maslow. *Motivation and Personality.* New York: Harper & Row, 1954.

5. Charles L. Hughes. *Goal Setting.* New York: American Management Association, 1965.

6. Peter Drucker. *The Practice of Management.* New York: Harper & Row, 1954.

7. George S. Odiorne. *Management By Objectives.* New York: Pitman Publishing, 1965.

8. W. Edwards Deming. *Out of the Crisis.* Cambridge, Mass.: Center for Advanced Engineering Study, Massachusetts Institute of Technology, 1986.

9. Kenneth H. Blanchard. and Spencer Johnson. *The One-Minute Manager.* New York: Morrow, 1982.

10. Thomas J. Peters and Robert H. Waterman. *In Search of Excellence.* New York: Harper & Row, 1982.

11. Robert R. Blake. and Jane S. Mouton. *The Managerial Grid.* New York: McGraw-Hill, 1964.

12. Michael E. McGill. *American Business and the Quick Fix.* New York: Henry Holt & Company, 1988.

Index